# PRACTICAL STRAW BALE BUILDING

To all of my family—a small token for very special people

# PRACTICAL
# STRAW BALE
# BUILDING

MURRAY HOLLIS

LAND
LINKS

National Library of Australia Cataloguing-in-Publication entry

Hollis, Murray.
Practical straw bale building.

Includes index.
ISBN 0 643 06977 1.

1. Straw bale houses. 2. House construction. 3. Building materials. I. Title.

693.997

Published by and available from
Landlinks Press
150 Oxford Street (PO Box 1139)
Collingwood VIC 3066
Australia

Telephone: +61 3 9662 7666
Local call: 1300 788 000 (Australia only)
Fax: +61 3 9662 7555
Email: publishing.sales@csiro.au
Web site: www.landlinks.com

Front cover
Photos by the author

Set in Sabon 11/14pt
Cover and text design by James Kelly
Typeset by Paul Dickenson
Printed in Australia by Ligare

# Contents

# Acknowledgments

I thank the numerous participants in workshops that I have conducted. Their willingness to share their diverse experience and contribute enthusiastically to the workshops inspired me to further effort and development of ideas. I thank the amateur and professional builders who have been happy to discuss their projects and experiences. I thank the many people around the world who have so willingly shared their ideas with others via the Internet. I thank scientists from various laboratories of the CSIRO who have freely shared their expert knowledge with me.

I thank Ken Anderson of CSIRO Manufacturing and Infrastructure Technology for providing me with the opportunity to undertake writing assignments, and Nick Alexander of CSIRO Publishing for providing me the opportunity to write this book.

I thank Jennifer and my children for accommodating my eccentricities and providing the support and encouragement that is so essential for production of such a work.

# 1
# Introduction

Although straw has been used for building for millennia—usually combined with other materials such as clay and sand—it seems that it was not until the baling machine was invented in the late 1800s that builders recognised the potential to use blocks of straw as a building material. Notably this occurred in Nebraska, USA, where traditional building materials such as timber and stone were not readily available. Some of those buildings from the late 1800s and early 1900s still exist.

A smattering of straw bale buildings was constructed up to about the 1980s, but the 1980s/1990s marks a substantial revival in straw bale building in many parts of the world, including North America, Europe and Australia. Now, examples of straw bale buildings include simple domestic dwellings of rather rustic nature, very modern homes, creatively sculptured structures and commercial buildings, such as wineries. They can be found in both urban and rural environments and, more often than not, their design emphasises other environmentally friendly features such as solar passive heating and use of other environmentally friendly materials.

Enthusiastic owner-builders are still building the large majority of straw bale buildings, although a small number of professional builders have adopted the material and are continuing to develop their methods. However, it would be surprising if many traditional builders, used to precision construction techniques, would embrace with enthusiasm a building material so unprecise and variable. Straw bales are awkward to handle; they are quite variable in length, shape, density, and surface finish—'finish' is hardly even an appropriate word in this context. Straw bales are also susceptible to water damage; they vary greatly in price and availability depending on the time of year, the weather in the growing season and the location; and they can create considerable mess and waste on the building site.

Straw bale house under construction at Jerrabomberra (Queanbeyan, NSW)

Without fundamental innovative changes building with straw bales will remain primarily in the alternative culture. To meet the challenge of making building with straw bales more attractive, or at least less abhorrent, to skilled building tradespeople, is one objective of this book. After an extensive discussion of the essential elements of current best practice in straw bale building—though many will debate what is best practice—the discussion ventures into innovate methods that should help to progress straw bale building technology—to move straw bale building further into the mainstream of the building industry.

This book does not address comprehensively all aspects of building a structure that has straw bale walls. Aspects of structures other than walls are addressed only to the extent that they are relevant to the use of straw bales as a building material for the walls. Issues not unique to building with straw bales, such as the various types of floors (concrete, earth, timber, etc.), roof structures, methods of heating/cooling, and sustainable building issues, are discussed extensively elsewhere in conjunction with straw bale building, as well as in other contexts. These issues would tend to dilute the main thrust of this book. However, some techniques not familiar to the building trade are discussed in substantial detail.

Some techniques are drawn from grain farmers, some from fence-builders, some from gardeners, some from the building trades, and some are new.

There are many environmental issues associated with straw bales. However, these are not covered in this book unless they are of particular practical relevance to the building technique being discussed. For example, many people claim that straw is a waste material and therefore environmentally ideal for building purposes. There is, no doubt, substantial truth in that claim, but the issues are many, and seldom are they fully explored.

There are many variations on the theme of straw bale building. Rather than attempt a comprehensive review, this book concentrates on techniques that are most likely to be accepted by the mainstream building industry. However, in numerous cases alternatives are mentioned, including techniques that have proven to be inappropriate or not very useful, but are still being applied.

Straw bale building has been incorporated into building codes, particularly in some states of the USA, often including some less-than-ideal methods. However, the published knowledge of straw bale building, including the material in this book, is insufficient to construct comprehensive building codes. Further research and more experience is desirable before regulatory authorities adopt very prescriptive straw bale building codes, though it would be desirable now for authorities to develop guideline documents that can be used in conjunction with the requirement that buildings meet specified performance requirements.

**Straw bales: more environmentally friendly than bricks?**

Manufacture of clay bricks begins with mining clay and other raw materials. The clay and other ingredients are transported to brick-making facilities where they are mixed, possibly screened, and formed into damp bricks. These are dried and baked to high temperature. They are stockpiled, transported to retailers, and finally transported to building sites—earthy, but non-renewable, and energy hungry. Compare that with straw, which is sewn, grown, cut, baled in the field, often stockpiled on-farm, and transported to building sites, usually bypassing the retailer. The absence of expensive energy-consuming factory processing and retail overheads, and straw being a renewable and recyclable material, tend to make straw bales economically and environmentally attractive.

# 2

# Basics of straw walls

Straw bale walls may be built by placing small rectangular bales, essentially bricks, in a 'running bond' fashion. The bales are not so rigid and precisely rectangular as bricks, so an unsupported wall of straw bales tends to be quite unstable. In fact, a wall more than about two metres high and a few metres long becomes rather like a slab of jelly, so temporary restraints are essential during construction.

However, when the stacked bales have been tied down to the foundations, thereby compressing the wall vertically, the wall is transformed from a mass of

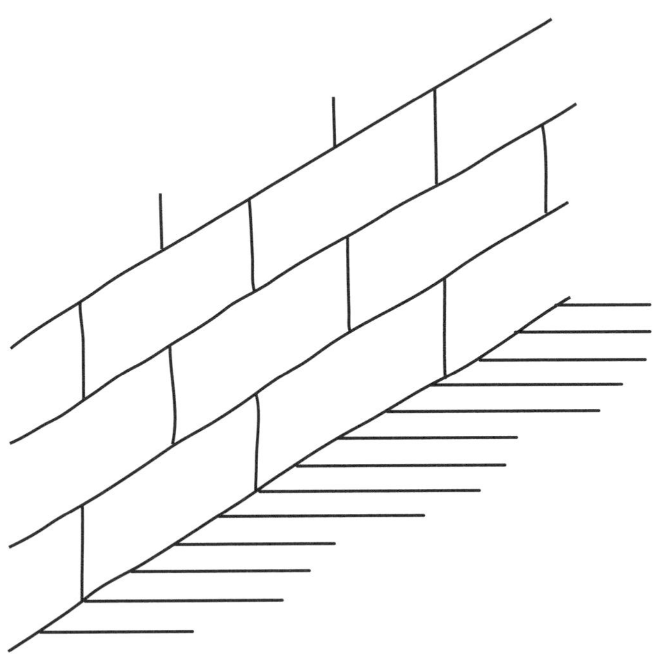

**Fig. 1** A running bond structure.

jelly to a stiff and resilient edifice. Usually straw walls are then plastered inside and out with three coats of lime-based, earthen or cement-based plaster.

The finished product is usually about a half-metre thick, with excellent thermal insulation characteristics, very good resistance to fire, able to support very substantial roof loads, and able to cope with strong impact—a secure, comfortable, durable, ecologically friendly and practical basis for buildings for domestic, commercial and farm purposes.

## Straw versus hay

Straw is the material that remains after a seed crop has been harvested. Hay is the finer grass, and normally is harvested as a feed material; for example, lucerne is harvested as hay, where the whole of the plant above ground is used for stock feed. For building, hay bales should be avoided. They are likely to contain more moisture, be composed of relatively soft material and usually will be more expensive.

Wheat, oats, rye, barley, and rice are some types of straw that may be used, and there are many other plants that can be used. As long as the straw is dry and can be formed into suitable bales, it should be suitable. Rice straw tends to have higher silica content, which might be some advantage, but it also tends to be relatively soft compared with oat straw, for example.

Of more importance than the type of straw is that it has fairly low moisture content (usually about 12% to 15%), and is baled in fairly uniform and fairly tight bales. If it is not green, appears to be dry, looks to be regular in shape and is well compacted, it is probably suitable.

## Straw bale sizes

Straw bales come in various sizes and shapes. The bales most used for building are the small bales that are approximately 350 mm × 450 mm × 900 mm. The common 'square baler' produces bales that are 14 inches × 18 inches (356 mm × 457 mm) and of variable length, which may be adjusted between about 300 mm and 1100 mm. Some older balers produce bales of 16 inches × 18 inches cross-section, but now these are not very common in Australia, having been superseded by the 14 × 18 inch bales for occupational safety (manual handling) reasons. In the USA there are also 3-tie bales, which are 14", 15" or 16" high by 23" deep by 43" to 47" long.

The largest dimension tends to vary a fair amount, because that is less easily controlled during the baling process, but the other dimensions are fairly consistent. Most producers now make bales that are much larger, either round bales (various sizes roughly 1 to 2 metres diameter), or large rectangular bales, which normally are either 800 mm × 900 mm or 900 mm × 1200 mm cross section, and of variable length, typically about 2000 mm. The latter have been used for large buildings, such as large winery buildings, but they require a forklift to handle them and, of course, take up a considerably larger area and require much wider footings. Use of these large bales is not considered in this book.

## The standard straw bale

The small rectangular bales have remained essentially unchanged for over a century. But just because farmers have found that 350 by 450 by 900 mm is a useful size of bale for storage and handling straw and hay, it does not follow necessarily that it is a good, or the best, size and shape for building purposes. But before analysing this issue, let us discuss the standard straw bale.

It is important to understand the structure of a straw bale, which is best done by learning how a straw bale is made—i.e. how a baler (a baling machine) works.

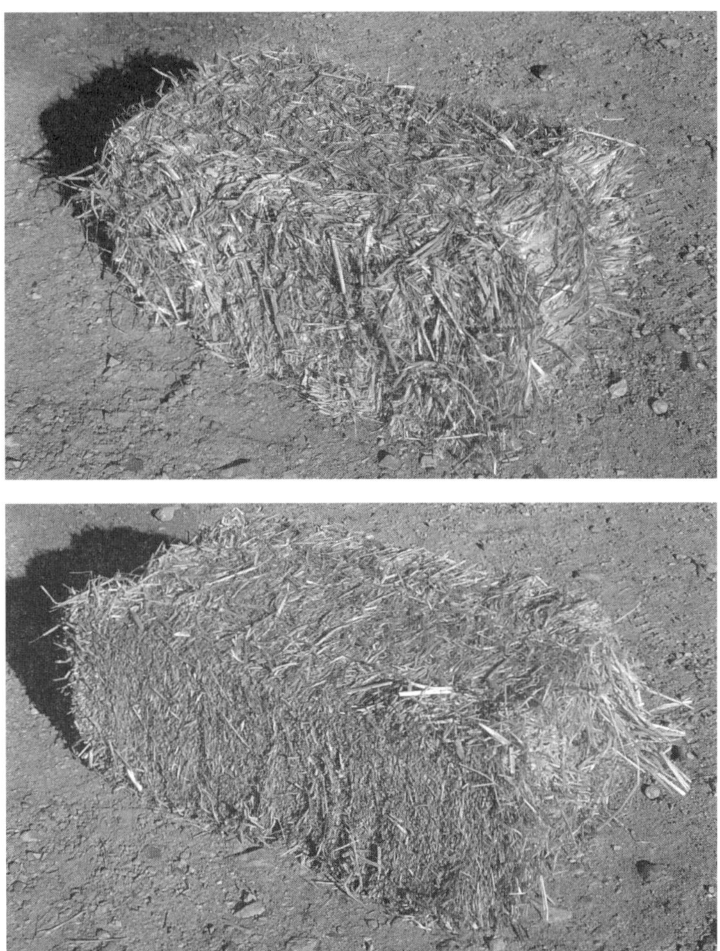

**Fig. 2** Standard small straw bale showing its folded face (top) and its cut face (bottom).

**Fig. 3** Baling machine in operation. The rotary rake picks up the straw, and steel fingers move the straw to the right. The ram that compresses the bale, the tying mechanism and the cutters are in the right hand section.

The straw is first cut and raked into rows in the field. The baler, pulled and powered by a tractor, straddles these rows and picks up the straw with a rotary rake. The straw is then transported sideways, usually with moving steel fingers, which push one end of the straw bundle against a stop—a flat steel face. In this process the straw is very roughly aligning in the direction perpendicular to the stop. The small bundle of straw that has been transported into position is then compressed with a moving plate driven by a hydraulic piston. This transportation of small bundles of straw and compression of the bundle is repeated in a pulsing action until the full bale has been built up. As the bale is being built up and pushed out of the baler, the face opposite the steel plate stop is trimmed (fairly roughly) with a cutter and the bale is mechanically tied lengthwise (usually two ties) with strong 'baling twine'. The bales can also be tied with wire, but this is not usual these days.

These actions result in the straws being very roughly aligned in one direction across the bale. Where the straw has been pushed against the flat steel plate much of the straw tends to be bent back on itself, and thereby forms what is called the 'folded face'. Because of the pulsing action the bale is composed of a number of fairly discrete 'biscuits' of straw.

These biscuits tend to be fairly discrete straw layers about 100 mm thick. The structure of biscuits is far from being a layer of straight parallel straws.

The straws are fairly randomly orientated and tend to be intertwined with only a general tendency to align in one direction. The straws are usually highly variable in length, but this depends very much on the nature of the crop and how it was mowed. At the folded face end of the biscuit the folding of the straws can extend about 100 mm into the biscuit. This folding of straws well into the bale can tend to cause that side of the bale to have a higher density. For that reason some people recommend that bales be laid with cut faces alternating on the two faces of the wall.

The end result is a bale that has a reasonably flat cut face—though, because of the nature of the cutter, that face usually has significant ridges across it—and a somewhat fuzzy folded face, which is far from ideal as a base for application of plaster (see Fig. 2).

The other four surfaces—the top, bottom and ends, as the bales are usually orientated—all have the straws roughly parallel to the surface. The top and bottom surfaces are fairly flat, but the ends can be quite distorted: the straws tend to bulge out around the baling twine on the ends.

A good understanding of these structural details of straw bales is very useful when one comes to place, tie-down, cut, trim and plaster them.

# 3

# Materials, components and tools

This chapter gives an overview of the main materials, components and tools you will need to build with straw bales.

## Tie-down wire

The strongly recommended practice is to tie down the walls to the foundations to make them stable and to minimise any tendency for them to settle over time. The best method devised to date uses 2.5 mm diameter (12.5 gauge) high tensile wire, which usually is available in 1500-metre rolls, more than enough for a typical domestic dwelling.

Numerous alternatives for wall compression can be found in the literature, ranging from nothing (let the walls settle for weeks or months before plastering—the 'Nebraska style') to threaded rods inserted centrally for the full height of the wall. Tie-down with high tensile wire tends to be far superior to published alternatives.

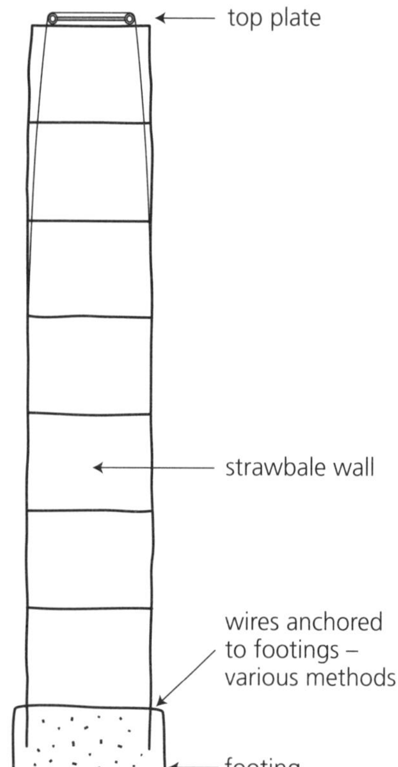


**Fig. 4** Wall tie-down geometry. Tie-down wires are anchored in the footings, pass over a top plate and are tensioned to increase the rigidity of the wall.

## Top plates

The compression wires are run over the top of the wall and down to the foundations. Since they are tightened to considerable tension, something must be placed on top of the wall to prevent the wires cutting into the top bales. This 'top plate' can take various forms. Most builders have used top plates in the form of heavy wooden ladders or wide planks of wood placed centrally on top of the wall. Even concrete beams have been advocated.

Such heavy-duty top plates are attractive for load-bearing walls, because they provide a rigid beam onto which the roof structure may be fixed. However, such facility is not required of the top plate in the case of in-fill walls. In fact, a much lighter, cheaper and more easily installed top plate can be used for both in-fill *and* load-bearing walls.

This alternative is a steel ladder that can be made from 16 mm to 25 mm nominal bore (NB) steel pipe; 25 mm NB pipe is the more appropriate size for load-bearing walls and will undergo less distortion for infill walls, which is particularly important when walls are tied down using the common fence strainer method, discussed later. The 'rungs' of the ladder may be 12 mm reinforcing bar or other relatively lightweight tube, pipe or rod; these rungs carry no significant load, since essentially they serve only to hold the pipes in position. However, for load-bearing walls, steel plates (say 50 mm × 10 mm section) may be welded between the pipes for attachment of the roof structure.

Such steel top plates should be no wider than 300 mm, and can be as narrow as 200 mm. If the ladder is close to the 450 mm width of the bales, it will be likely to slip over the edge of the wall in some places during the tensioning of the wires.

As the wires are tensioned they will cut into edge of the top bales, but rather than being a disadvantage, this is desirable, because thereby the wire tension is applied more centrally to the wall, reducing the tendency for this force to cause the wall to tilt.

These steel top plates can be fabricated very quickly on-site as needed using an electric welder and an angle grinder or hacksaw. (Be sure to keep the welding and angle grinders well away from any straw bales or loose straw, and have water or fire extinguishers handy in case of fire.) The materials are relatively cheap, especially if one uses downgraded pipe, which might have small defects that do not significantly affect its strength. Remember, this top plate will be completely buried in the final wall, so if it has surface rust or has some unattractive surface finish, that will be of no consequence.

Another significant advantage of steel top plates is that they may be fabricated quite readily for curved walls (see Chapter 12 on creative shapes).

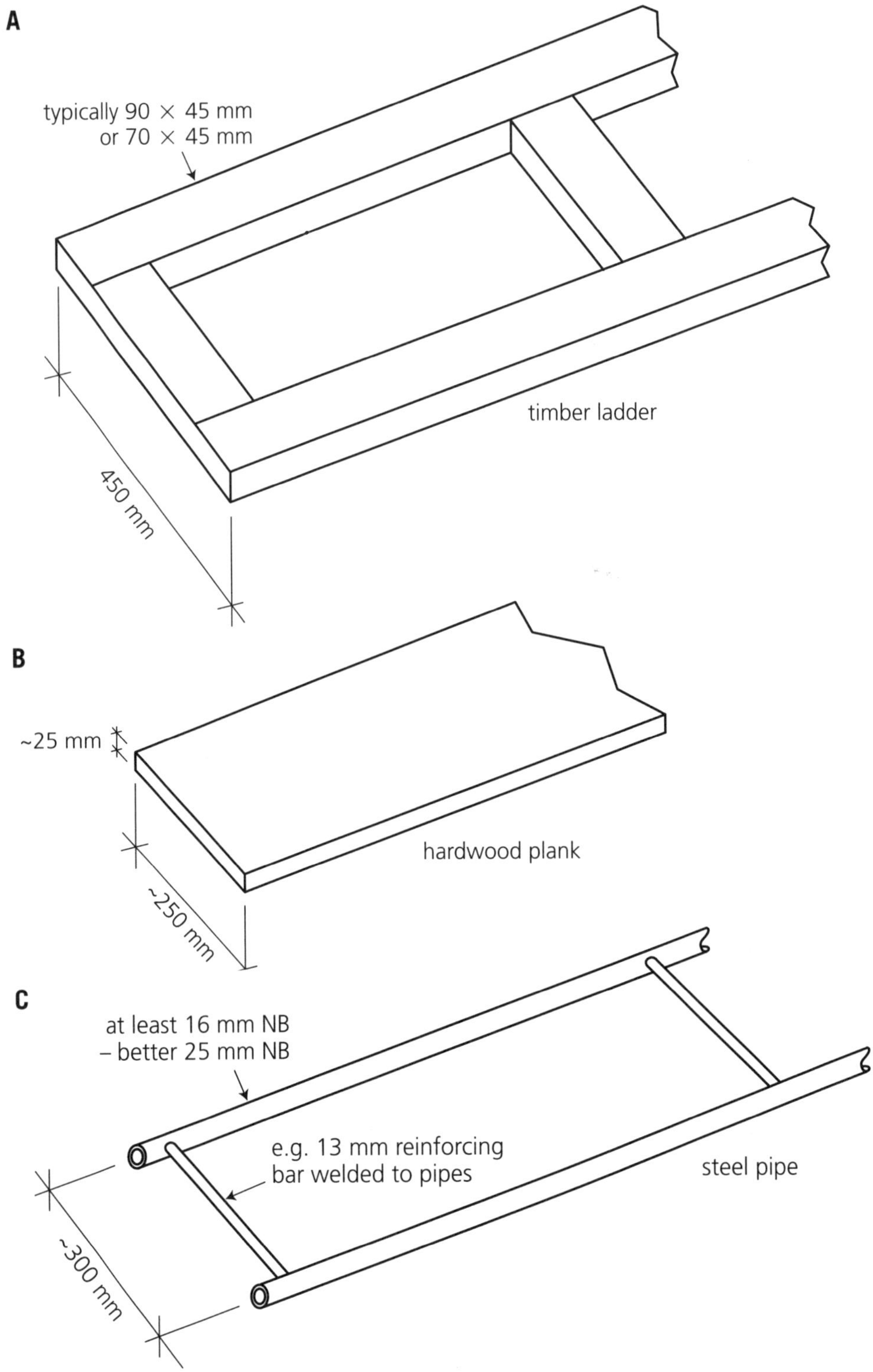


**Fig. 5** Top plate alternatives. Top plates may be of wood or steel. A: Timber ladder top plate. B: Hardwood plank top plate. C: Steel pipe top plate. D: Modified steel pipe top plate for load bearing walls.

**D**

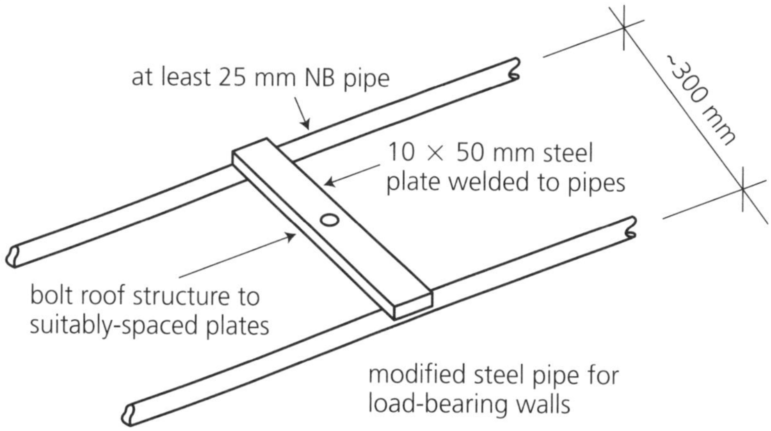


## Intermediate structural units

Structural units somewhat similar to top plates may be used at intermediate stages of wall construction, particularly as an aid in constructing high walls, and to provide anchorage for anything from shelving to artwork to drywall cladding. These units may be of simple, lightweight steel ladder structure, with the addition of strips of metal or wood at the wall surface (beneath the plaster) if they are to be used for anchorage. If they are only to assist in the construction of high walls, they can be a simple steel ladder structure, as described above.

Intermediate structural units to be used for anchorage of drywall, shelving, etc. may be made from trench mesh that is the same width as the bales (usually 450 mm) with lightweight steel angle (such as 51 mm × 30 mm × 3 mm galvanised angle) tack-welded to the sides of the mesh. Since 450 mm wide trench mesh generally is not available, if the full width is required (i.e. for anchorage on both wall faces) it has to be fabricated from reinforcing bar. However, if anchorage is required only on one face, then 300 mm wide trench mesh may be used, with the steel angle fixed to only one edge. For the 300 mm wide straw bales (to be introduced later—see page 64) these units can be made using standard 300 mm wide trench mesh and the same lightweight steel angle, or 200 mm trench mesh for single-side anchorage.

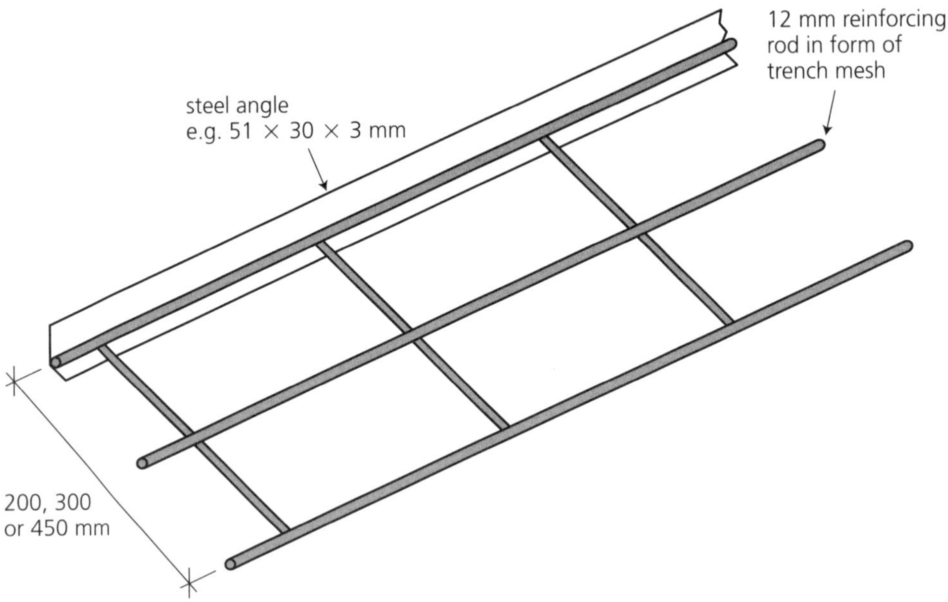


**Fig. 6** A design for intermediate structure units for anchorage of shelving, drywall cladding, etc. The steel angle may be welded to one or both edges, as required, but usually it would be on the inside edge only.

## Wire locks

Wire locks are devices that enable wires to be joined whilst allowing the wire to be re-tensioned. Available wire locks include the Wirelok® made by Wirelok Ltd (NZ), and the Gripple® made by Gripple Ltd (UK). They have slightly different mechanisms but are similar in function.

The wire locks have two holes that accept the wires but allow each wire to move only one way through the holes. Wire locks are used primarily for rural fencing as an alternative to using wire knots. It is not feasible to use knots for joining tie-down wires for straw bale walls, because every wire must by re-tensioned, usually a number of times. This is not possible with the wire knots.

However, there might be occasions when two pieces of wire have to be joined in a permanent, non-adjustable manner (e.g. a wire has been cut too short and it is not convenient to start over again). In such cases, an effective and very simple knot, which can be used with high tensile wire, is the figure-of-eight knot illustrated.

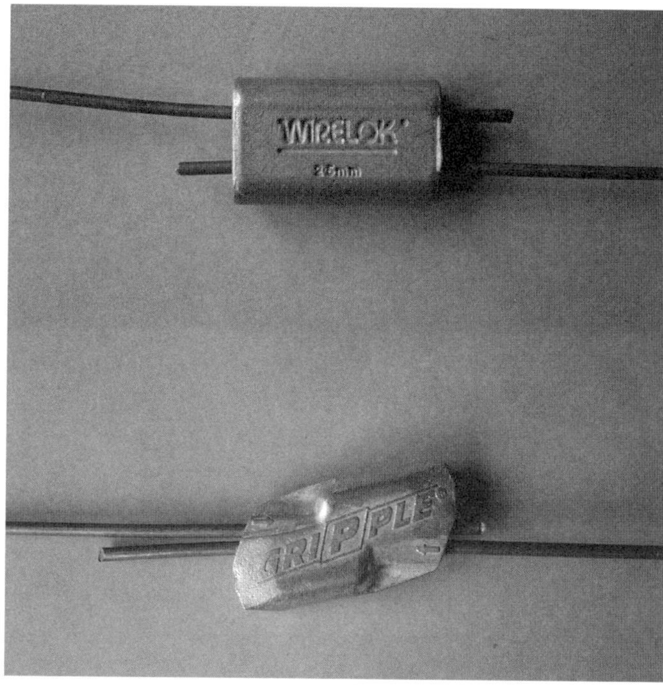


**Fig. 7** Wire locks for joining 2.5 mm diameter high tensile wire. The Wirelok® (top) is made by Wirelok Ltd (NZ) and the Gripple® (bottom) is made by Gripple Ltd (UK).

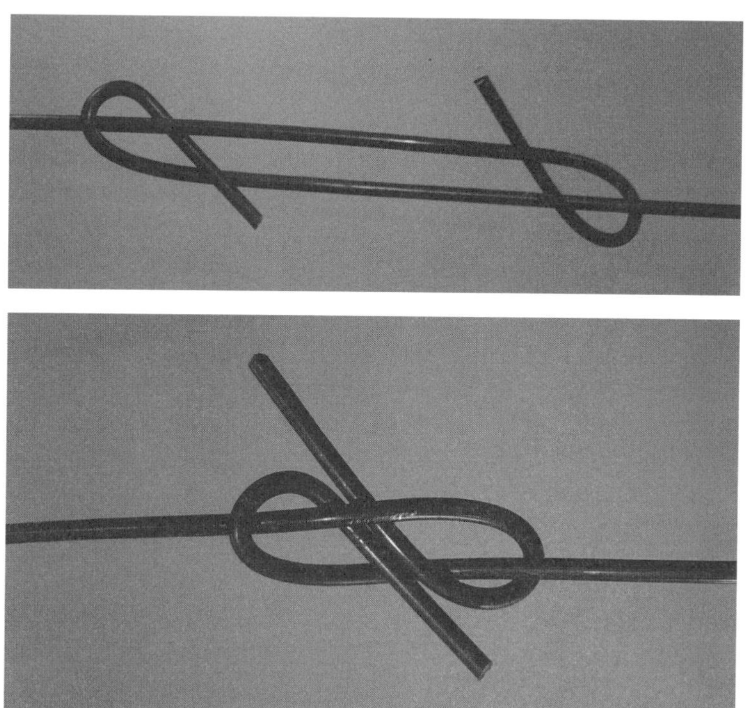

**Fig. 8** A figure-of-eight knot for joining high tensile wire. Make the knot as shown (top), then pull together (bottom).

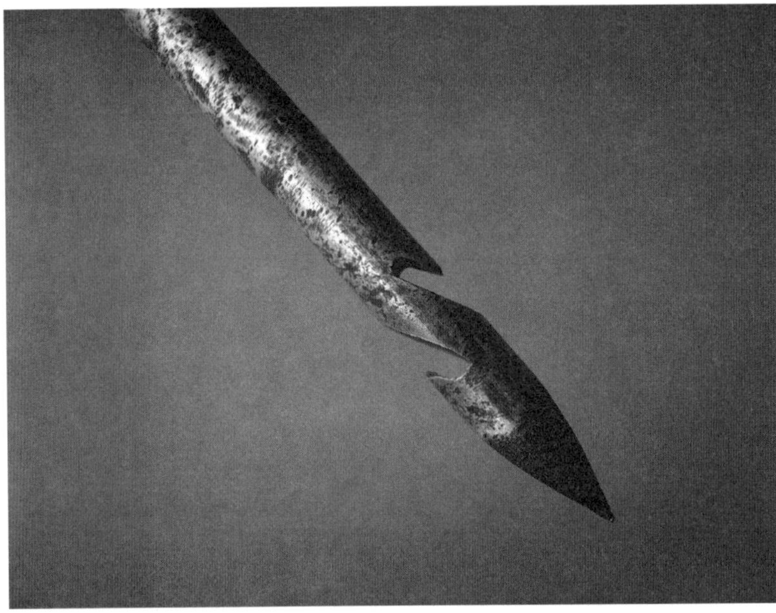

**Fig. 9** Needle used for threading twine through a bale when splitting bales. Twine may be inserted by both a push and pull action.

## Needles

Special large needles are very useful when it comes to splitting straw bales, which is usually referred to loosely as 'cutting' the bales, though the bales are not actually 'cut' in most cases. The needle should be about 500 mm long so it can be passed easily through the 350 mm dimension, and also used through the 450 mm dimension in case that becomes necessary. It should be made with two slots to carry twine through the bale by a push—for one piece of twine—and then a pull action—for a second piece of twine. Holes may be used instead slots, but it is much quicker to drop the twine into a slot than to thread it through a hole.

These needles can be made from 8 mm diameter mild steel rod with a short piece of the same welded on one end to make a handle, or simply a bent end to form the handle. If the rod is smaller than 8 mm diameter it can become too weak at the slots. The slots can be cut easily with a cutting disc on an angle grinder.

## Push-pull extractor

The 'push-pull extractor' can be useful if a deep hole is required in, or right through, an un-plastered wall, such as to install plumbing, though it is much better to install plumbing as the wall is being built. This tool is a somewhat fearsome device and requires some basic welding skills to make. An angle

**Fig. 10** A 'push-pull extractor' tool, used for extracting straw from a wall to produce holes, such as for pipes. It may be made by welding masonry nails to a steel rod, with a suitable handle on the other end.

grinder is also very useful. Near the end of an 8 or 10 mm diameter mild steel rod, weld reverse-facing barbs made from heavy-duty nails, at least 3 mm diameter, and sharpen the end of the rod to a point. Masonry nails are ideal as barbs, because they are hardened and therefore less likely to bend in use.

This tool may be pushed into a bale quite readily, and then, as the tool is withdrawn, the barbs drag out straw. A rapid push-pull action enables the removal of straw at a reasonable rate, though it still requires considerable effort to make a hole right through a wall.

## Wire tensioners

A wire tensioner is necessary to tension wires for holding down and compressing the walls.

### Traditional fence strainer

Some practice is required to become a proficient user of the common type of fence strainer (also called a chain pull) used for rural fencing, but these tensioners are quite suitable, though not ideal, for tensioning wires on straw bale walls. The main constraint is that the wire lock join should not be placed very close to the bottom or top of the wall, because for this type of wire strainer a substantial clear length of wire is required for easy operation of the strainer.

**Fig. 11** Using a traditional fence strainer requires a bit of practice.

This issue usually is of no concern for fencing, but it is significant in the more restricted space of straw bale building. The fairly cumbersome operation of these wire strainers, especially in confined spaces, makes somewhat inefficient the repeated attachment of the strainer during re-tensioning wires.

## Pop riveter as a strainer

In the case of the Wirelok®, wires may be tensioned using a pop-riveter gun, which is very convenient in confined spaces. The pop-riveter gun enables only a small movement of wire for each operation of the gun, but the action can be repeated very quickly, so this is not a significant disadvantage. However, a pop-riveter gun cannot be used with Gripples®, because the holes for the two wires are too close together, leaving insufficient room to attach the gun.

## Gripple® strainer

There is a strainer specifically designed for use with Gripples®. It overcomes the space problem and can be attached much more easily and quickly than the traditional fence strainer. In this case, and for the pop-riveter gun, to enable re-tensioning of the wires, the waste wire must not be cut very close to the wire lock, because these strainers use this protruding piece of wire.

**Fig. 12** A wire spinner is essential for working with rolls of high tensile wire to prevent horrendous tangles.

## New wrench strainer

I have developed a new strainer specifically for tensioning wires on straw bale walls. It overcomes the access problems, removes a problem due to friction between the wires and the top plate, enables very simple, accurate and quick re-tensioning of wires and requires only a standard wrench to operate. It forms part of a new straw-bale-wall construction system (see Chapter 8).

## Wire spinner

Since the high tensile wire usually is supplied in 1500 metre rolls and is very springy, when an attempt is made to unwind the wire it will tend to spring in every direction and a horrendous tangle is the likely consequence. Therefore a wire spinner (available at rural suppliers) is essential for handling these rolls.

## Wire cutters

Many ordinary pliers and wire cutters do not have long enough handles to provide the leverage necessary for easy cutting of high tensile wire. Mini bolt cutters with 300 mm-long handles and hardened cutting edges will do the job more easily, and with greater safety for the tool and the operator.

> **Safety warning**
>
> The high tensile wire is very springy and therefore can easily flick into the user's eyes and cause serious damage. Take great care when handling this wire. For example, do not leave wire ends hanging off walls. If it is not convenient or desirable to cut such wire ends, they can be poked into the straw wall to minimise the chance of injury.

## Staples

There are many references in the literature to 'stitching' wire mesh or metal lath to walls by threading twine or tie-wire through the wall so it may be used to pull wire mesh hard against the wall. There are also many references to stretching wire mesh across the face of the wall and fixing its edges, to timber for example, to enable the wire mesh to be held in tension.

Quite apart from the fact that wire mesh or metal lath is not necessary in most situations, if it is used, generally it is not necessary to stretch it or hold it in tension, and the stitching procedure is very labour intensive compared with the use of staples. If the straw face is not a good cut face, stretching wire mesh across the face can make plastering easier, but if the straw has a good cut face, stapling the wire mesh to the wall face is entirely adequate.

Staples can be made from the same high tensile wire as is used for compressing the walls by cutting 500 mm lengths of the wire and bending them into a 'U' shape. These can be used to hold wire mesh in place simply by pushing the staples through the mesh into the wall. They can also be used for such purposes as holding electrical plaster plates in place (see page 83), and to secure temporarily the edges of plastic sheet, as on the top of garden walls (see page 88).

## Mallet

A large wooden mallet is almost essential for straightening straw bale walls that are built by the usual current methods, but a new method (see Chapter 8) eliminates the need for such mallets.

## Brush cutter/cord or string trimmer

Straw bales do not lend themselves to trimming and shaping by the usual carpentry and building tools. There are a number of tools that can be used for this purpose, including: hedge cutters, cord trimmers, chain saws and brush cutters.

Chain saws are not very suitable. Apart from their being particularly dangerous in inexperienced hands, the cut straw can easily enter the drive

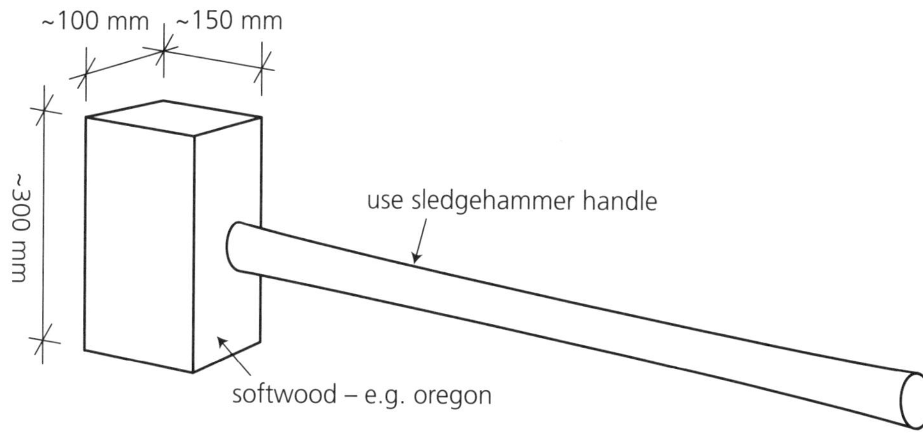


**Fig. 13** A large wooden mallet should be used to straighten straw bale walls.

mechanism, and cause it to jam, which usually requires the saw to be dismantled to remedy.

Power hedge trimmers can do a reasonable job, but are not as versatile as the other alternatives.

Nylon cord trimmers are the most commonly used trimmers for this purpose, but brush cutters are a very good alternative. Brush cutters are usually supplied with rigid cutter discs, but also available are cutters with pivoting blades (e.g. the Weed Wakka®), which are more forgiving when they hit solid objects and tend to be longer lasting.

## Bale compressor

A device for compressing bales is not an essential tool for straw bale building, but it can be useful for binding existing bales more tightly and to recover bales when the twine has been cut or broken (e.g. been eaten by rodents).

The bale compressor shown in Figure 14 was made from scrap steel. The incentive for designing this bale compressor was that rodents, which seem to have a liking for the twine, had attacked a stack of bales in storage. Most of the straw was intact, but most of the ties were broken.

This compressor has a mechanical advantage of approximately 10:1, so a 30 kg force applied by the operator will apply about 300 kg force to the bale, which is adequate to achieve good bale compression.

It should be feasible to use a modification of this device to manufacture bales from scratch, which might be attractive in regions where straw of some variety is plentiful, and so is labour, but expensive mechanised baling machines are not available due to poor economic conditions—i.e. in some developing countries.

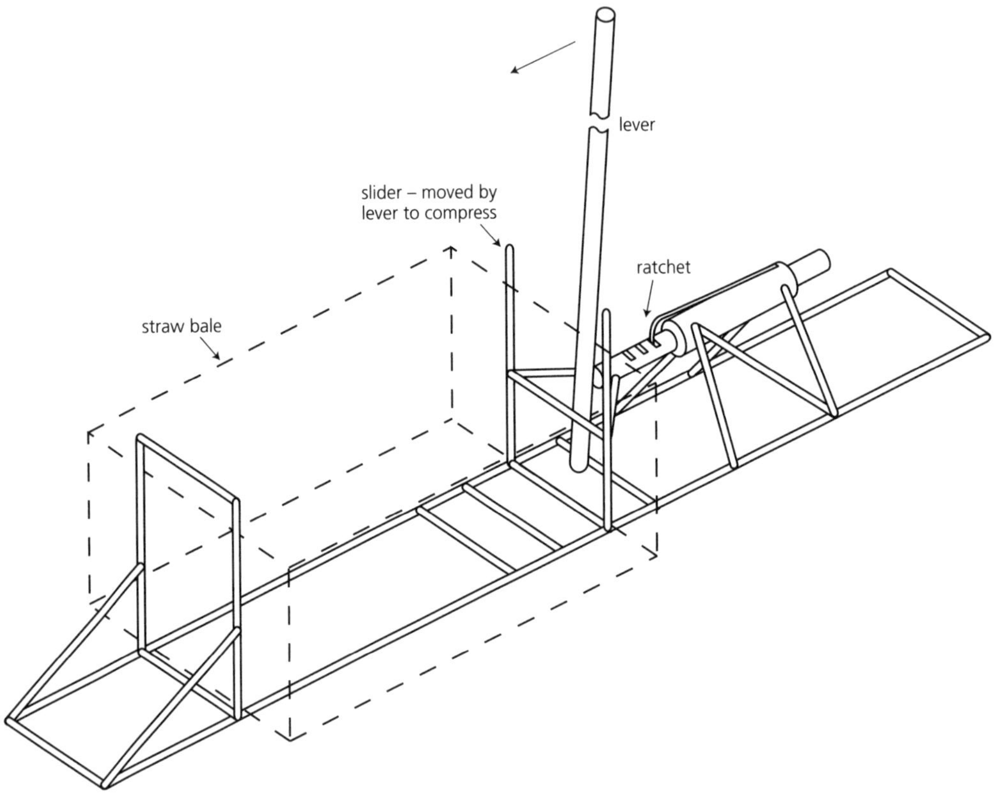


**Fig. 14** A bale compressor may be used to recover defective bales, particularly bales with broken ties.

# 4

# Preparing the bales

In addition to the following processes for preparing bales, there are other ways of shaping bales that are discussed in Chapter 12.

## Bending and straightening bales

Bales can be bent readily to suit curved wall construction, or straightened if they have been inadvertently bent during handling. Always they should be checked for straightness before incorporating them into a straight wall. Bending is easily achieved by placing the bale edge-down with one end on a chock of some kind and applying force with one's knee.

## Pre-trimming the bales

The folded faces of the bales should be trimmed before placing them in the wall, because the cut faces are much easier to plaster and they are more easily cut before, rather than after, erection; after erection the tie-down wires interfere with trimming. Bales with neatly trimmed faces are also more easily aligned to form a straight, flat wall.

To trim the folded faces, lay bales folded-face-up on a flat surface. The bales can be placed in a layer as large as desired, but it is best to have the width small enough that the centre of the layer can be reached without walking on the bales.

Trim the bales enough to produce a cleanly cut face, and keep the bales to a uniform thickness with the faces as flat as reasonably possible. This trimming will reduce the thickness of bales, but usually only by about 20 mm, which can be accommodated readily.

## Splitting bales

To make a part-bale, the bale is not actually ‘cut’; the bale is re-tied, the original ties cut, and the bale separated into two pieces.

Note the reference to ‘part-bales’, not ‘half-bales’, because each part-bale should be tailored to fit the space available (see page 35).

Thread baling twine through the bale using the bale needle at the desired new bale length. If both pieces of bale are to be preserved, insert one piece of twine in the first pass of the needle, and another piece on the withdrawal of the needle, but take care not to overlap the two pieces of twine. Practice is the best method of learning the technique, but essentially:

- the slots on the needle should point towards the ends, not the sides, of the bale;
- after inserting the first piece of twine (on needle entry), lay the ends of the twine towards one end of the bale;
- similarly lay the ends of the second piece of twine (on needle extraction) towards the other end of the bale;
- immediately proceed to tie off one, then the other piece of twine around the bale.

When all the new ties are in place for each shortened piece of the bale, cut the original ties and pull the bale apart. The new bale ends will be somewhat ragged, and therefore some trimming might be desirable.

An easy and effective knot to use for tying the bale twine is a ‘butchers’ or ‘packers’ knot, of which there is more than one variety. The simplest of these is formed by taking one end of the twine and tying a simple half hitch over the other end, and then pulling the second end through that half hitch and back on itself. The half hitch tightens and grips the second end as the second end is pulled and thereby slippage tends to be minimal. While tension is still maintained, tie off the second end around the twine, twice. The twine might slip around the bale as tension is applied on the second end, but this can be minimised by putting a foot or knee on the twine around the bale.

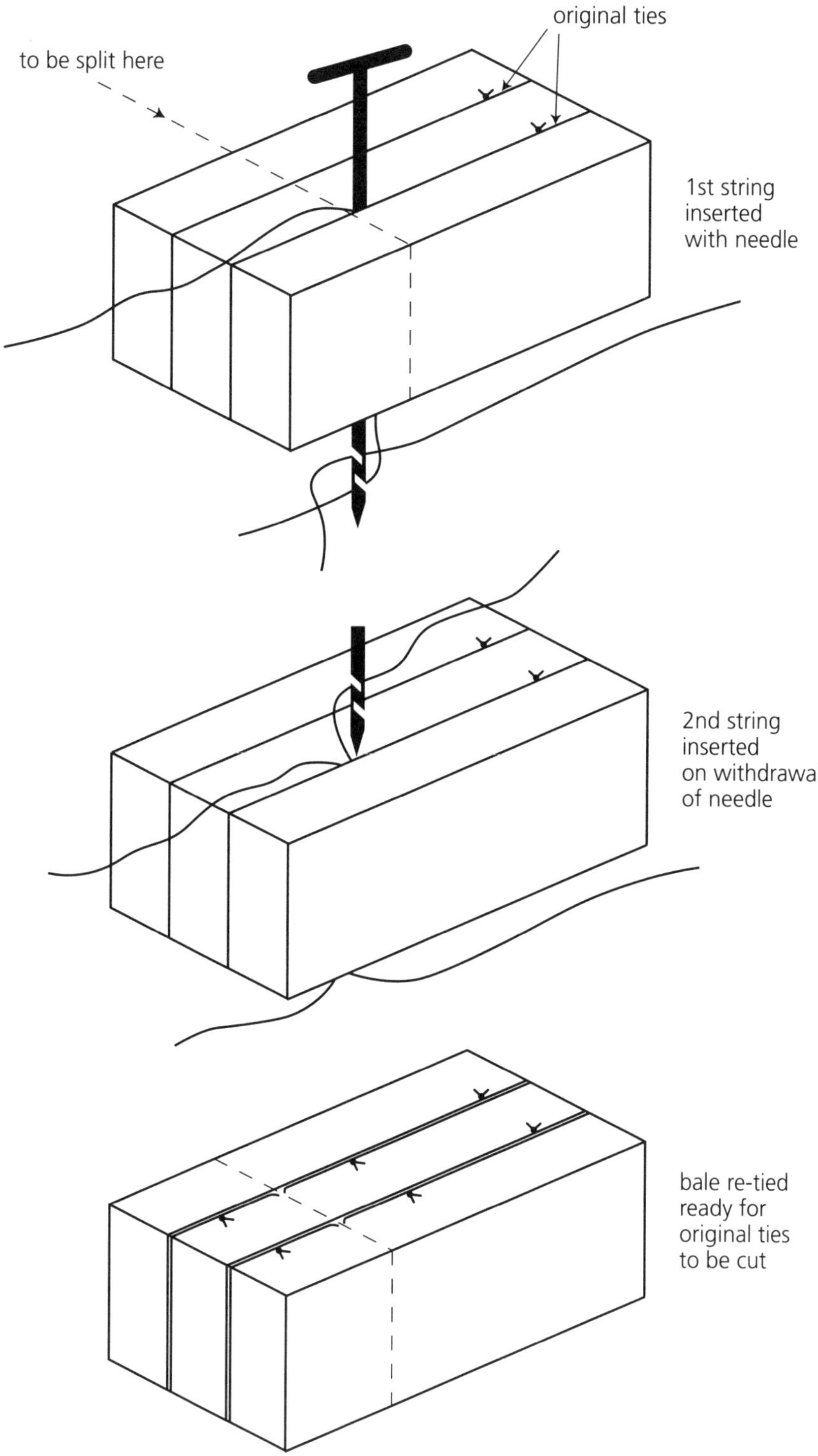


**Fig. 15** Procedure for making part-bales, by splitting and re-tying bales. A: the first string is inserted with the needle. B: the second string is inserted on withdrawal of the needle. C: the bale is re-tied before the original ties are cut.

# 5

# Foundations

Buildings need foundations. Straw bale walls are quite heavy and quite wide—about 500 mm—so the foundations need to be strong, and wider than usual. A 2.4 metre high wall made from 450 mm thick straw bales with say 30 mm thickness of plaster on both sides will weigh approximately half a tonne per metre length of wall, or the wall mass will be approximately 230 kg per square metre of wall. The straw bale component of this is approximately 65 kg per square metre of wall, but the plaster component is about 180 kg per square metre of wall (two faces). Clearly, plaster is the major contributor to the weight of the wall. Every 12 mm thickness of plaster (two faces) weighs roughly the same as the straw in the same area of wall.

A number of writers have stated that it is important that the plaster be supported directly by the footings, rather than allowing the force due to that mass to be transmitted through the bales to the footings. That is, the width of the footings should be sufficient to span the full width of the finished wall. Although it is probably wise to apply this guideline, one might question whether a 500 mm-wide footing really is necessary. Consider a 450 mm-wide footing (same width as the bales), so the plaster does not bear directly on the footings, but is supported only by its adherence to the straw. For a 2.5 metre-high wall and 25 mm thick plaster, the plaster mass to be supported is approximately 350 kg per metre length of wall (two sides), or 70 kg/m$^2$ on each side. This 350 kg compares with at least 1000 kilograms force loading per metre of wall by the tie-down wires (using guidelines as detailed on page 38). These figures suggest that it might not be necessary to use the wider footing, but probably it is wise to do so until empirical testing proves they are not necessary.

Some innovative schemes have been devised for wall foundations, but the essential requirement is that they must be sufficiently strong and stable to carry the load of the wall plus the roof structure and to transfer that load to the underlying earth in a manner that ensures that the building remains stable in all foreseeable environmental and building use conditions. That might seem to be of limited practical help, but that is fine, because engineering advice should be obtained about the specification of the foundation requirements for the particular earth and building conditions that apply; structural engineers are qualified to deal with all these issues, so this discussion will outline only the foundation requirements.*

Typically foundations are a concrete mass to the full width of the wall (about 500 mm), roughly 300 mm deep (dependent on the earth conditions) and reinforced with appropriately placed 'trench mesh'. There may be additional requirements for footings beneath columns, and there are detailed matters to be considered of how the footings link with slab floors and other building elements. These issues are not peculiar to straw bale building, and therefore are not dealt with here.

The essential difference between footings for straw bale walls and other footings are:

- they have to be wider than usual, about 500 mm;
- either the top of the footings should be substantially above the surrounding ground and floor levels or some intermediate element should be placed between the footings and the first layer of bales as a precaution against flood water permeating into the bales; and
- special provision may be incorporated into the footings to facilitate the attachment of wall tie-down wires.

The use of footings other than concrete is one that relates to issues such as personal preference, environmental considerations, and materials availability, and therefore is not within the scope of this book. Such considerations are relevant to many other building techniques as well as straw bale building.

Footings should be at least 150 mm above surrounding ground levels and at least 50 mm above the adjacent floor level. Special considerations may be necessary on sloping sites. Although the plaster might well prevent floodwater entry, it is much better not to rely on that. Foundations for a strongly sloping site may comprise columns of concrete, steel or treated timber that support beams on which straw bale walls are constructed, or special attention to drainage, especially storm drains, may be necessary.

If the raised section is not poured integrally with the rest of the foundations, typically the raised section may be masonry or timber. Options include ordinary house bricks and hardwood (e.g. 90 mm × 70 mm).

* See also Australian Standard AS2870 Residential Slabs & Footings

Some factors to consider are:

- what are the foreseeable ways in which flooding could occur, taking into account potential future changes? This will help to determine the step height to use;
- termites could attack timber, so that might suggest avoidance of timber (see below);
- plaster will be applied down to the footing, and will adhere better to bricks than to timber; and
- bricks, timber or other material used need not be of good appearance, because it will be buried under plaster in due course; however, it must have the necessary structural integrity.

These considerations tend to tip the balance in favour of bricks. Placed edge-on, with 10 mm of mortar below, the step height will be about 120 mm, a reasonable compromise; placed flat it will be about 90 mm.

For a timber step, the choice of timber type will be determined primarily by what is available readily. Used timber, including timber with nails still in it, or slightly twisted timber is likely to suit in most cases. Again, note that it will be buried, and its only purpose is to provide a step for the wall to sit on.

The space within the frame that forms the bottom unit of the wall should be filled with loose material such a coarse aggregate. This serves two purposes: it provides a flat surface to support the bale wall; and by choosing a coarse material such as 10 mm clean aggregate, wicking of water up to the bales by capillary action is minimised.

Although the aggregate should prevent damp rising by capillary action, a dampcourse should be placed before placing the first bales. This is secured simply by the weight of the first layer of bales. Dampcourse is substantially thicker than the plastic underlay used under concrete slabs. It is available in various widths, including 450 mm.

Dampness is also a hazard for timber floors, where good air movement under the floor is important to prevent the timber rotting. Since straw bale walls are usually much thicker than standard walls, if timber floors are used, the sub-floor ventilation openings should be made substantially larger than usual.

## Making provision in footings for tie-down wires

At the footing stage one should consider whether to incorporate facility for threading tie-down wires through the footing. If this method is to be used, appropriate tubes should be inserted into the footing. Alternatives include 13 mm plastic pipe ('poly pipe') and 20 mm diameter electrical conduit (only suitable for straight-through use), but one may use any plastic tube that can be readily bent to the required curvature, is rigid enough that it will not easily squash during pouring footings, and is at least 7 mm inside diameter (to accept at least two wires of 2.5 mm diameter). If using low-density 13 mm diameter poly pipe, as used for garden irrigation, care must be taken that it does not bend or squash during pouring the footings, because it will be very inconvenient if the tubes become blocked. The ends of the tubes do not need to be treated in any special way, such as bending them up, because modern high-tensile fencing wire will cope with sharp bends.

For cases where a curved tube is necessary, the 13 mm poly pipe is a reasonable choice, because it is very easily bent into the curved shape—but note the caution mentioned above.

To avoid the tube being filled with concrete as the footing is poured, and to make it easy to find the tubes later, use a tube length that extends well beyond the edge of the footing. A cloth or paper bung can be inserted in the tube ends or the tube can be bent back on itself and clamped as an extra precaution, but usually that is not necessary.

The footing dimensions are important. Each pair of tie-down wires should leave the footing at a separation of 450 mm, but if the footing is 500 mm-wide one needs either a narrower section on the upper part of the footing or use of a curved tube.

An alternative to installing tubes in the footings is to incorporate tubes in a separate raised section of the footings.

If wire-threading tubes are used in the footings, there is no need to fix the timber to the footings. The timber can be constructed as a ladder and simply laid in place. However, if the tie-down wires will be attached to the timber step, rather than being inserted into the footings, the timber must be secured to the footing, for example with masonry bolts.

If the footings are already in place without tubes inserted for tie-down wires, provision needs to be made for dealing with the tie-down wires. There are numerous alternatives and all are more complex than using the tubes in footings.

## Tying to a timber step

If wires are to be secured to timber at the wall base, the timber should be secured to the footing, usually with masonry bolts. When the timber is so fixed to the footings, a ladder structure is not necessary; a pair of parallel rails should be adequate.

Holes alone through the timber to accept tie-down wires are not adequate, because it will not be possible to thread the wire after the aggregate and bales are in place. Therefore, tubes—e.g. plastic electrical conduit, agricultural 'poly pipe' or metal tube—must be inserted through the timber rails. These should be inserted at mid-level or lower, to guard against timber damage when the wires are tensioned. If this method is used, hardwood should be used to achieve the necessary strength to cope with the force exerted by the wires on the wood.

## Using tubes bolted to the footings

Tubes may be bolted to the footings before a timber or brick step is put in place. In this case a timber step need not be bolted to the footings, and the method can be applied whether the step is of timber, brick or other material.

Note that wire-threading tubes for tie-down cannot be incorporated safely under a run of bricks, because when tension is applied the bricks are likely to lift. Therefore, wire-threading tubes either should be inserted in the footings or clamped to the footings in the case of using bricks for the raised section of footings.

## Anchoring tie-down wires directly to footings

Tie-down wires may be tied to an anchor (e.g. eyebolt) fixed to the footing on each side of the wall. In this case one might use a wire knot, such as is used in fencing, but such knots are not easy to tie using high tensile wire. Alternatively, a wire lock can be used to secure wire threaded through an eyebolt. The anchor needs to be right at the edge of the bale, but it is not desirable to attach wires to the anchor before the bales are laid, because the wires then become a hazard during bale laying. However, after bales are laid the anchor points can be somewhat obscured by the bales. Attaching wires to the anchors using wire locks is not too difficult, but attachment using wire knots can be very difficult.

Clearly, tubes-in-footings is the simplest and most economical method, the next choice probably being either tubes in a timber base, or steel tubes bolted to the footings.

**Fig 16** (right) Options for the placement of tubes in footings for tie-down wires. A: The footing is poured with a step section as an integral part; the step rises well above the floor and surrounding ground level. The floor slab at right may be poured at the same time as the footing. A tube is laid through at floor level before the concrete is poured, and the excess part of the tube is removed just before inserting the tie-down wire. Note that this option does not include facility for the plaster to bear directly on the footing. B: This option is as for A, except the width of the lower part of the footing is increased to enable the plaster to bear directly on the footing. C: If a separate wood or masonry step section is to be added after the footing has been laid, a curved tube may be incorporated into the footing when it is poured. In this case the step section does not need to be secured to the footing, because it will be held in place when the tie-down wires are tensioned. D: If no provision is made for tubes in the poured footing, the tubes may be incorporated in a timber step section, but this requires that the timber be bolted to the footing.

**A**

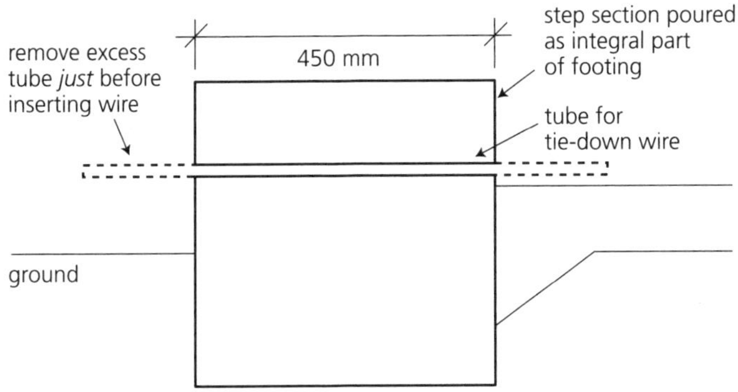
step section poured
as integral part
of footing
450 mm
remove excess
tube *just* before
inserting wire
tube for
tie-down wire
ground

**B**

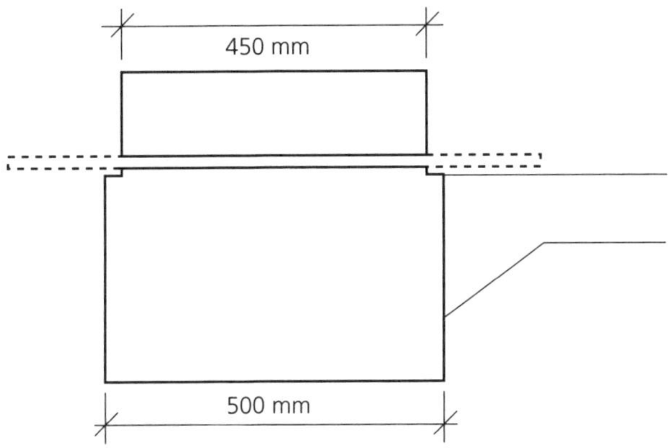
450 mm
500 mm

**C**

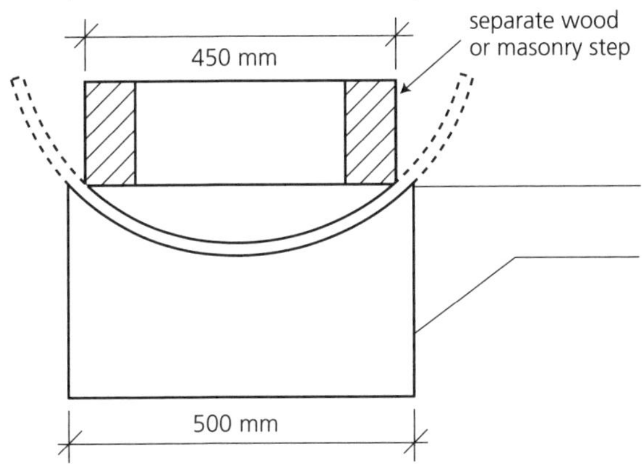
separate wood
or masonry step
450 mm
500 mm

**D**

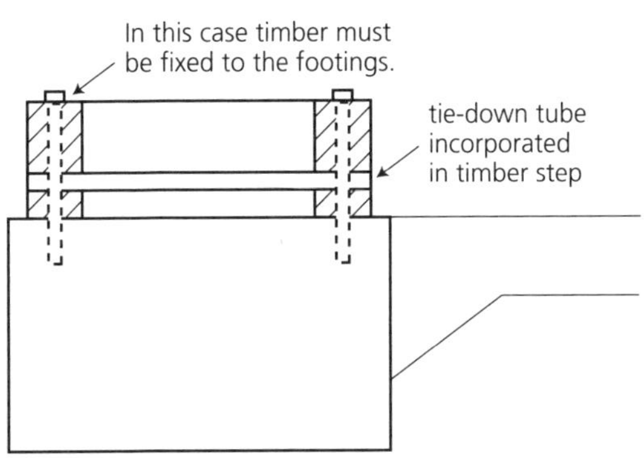
In this case timber must
be fixed to the footings.
tie-down tube
incorporated
in timber step

## Termites

There appears to be no reported cases of termites eating straw in straw bale walls, but it also appears that no experiments have been conducted specifically to test for this potential problem. There are many types of termites and there are a limited number of straw bale buildings. Therefore, the absence of reported damage from termites does not provide good reason to ignore this potential problem.

Known information includes:

- some termites, such as 'harvester' termites, which are active in some drier natural pasture areas of Australia, do eat grasses, though there appears to be no evidence of them eating crops such as wheat;
- termites sometimes chew material other than wood to construct tunnels to access across exposed surfaces but this is not likely to be problem in itself; and
- there is a very large number of termite species worldwide, so there is a lot of room for research into their interest in straw walls.

Even though the termites might not be interested in the straw, it is possible that they could travel through a straw wall, including through earthen plaster or through fine cracks in other types of plaster, to reach wood that is within the wall, such as at its base, as is common, or to reach wood within the building. This is little different from the situation with other types of building.

Therefore, although there is a lack of evidence of termite problems with straw bale walls, it is wise to take normal precautions to minimise the chance of termites entering the walls. Simple measures include keeping timber off the ground in the near vicinity of the building, and providing a viewing strip of concrete footings that is kept clear of all growth and debris and can be regularly inspected for evidence of termite tunnels. Barriers that may be used to deter termites are described in the Australian Standard AS 3660.1.

**A note about 'starter bars' and stakes**

This note concerns practices that are *not* recommended, except in one special type of case (see below). However, since some of them have been promoted as highly desirable features of straw bale building, and often presumed or implied to be essential—some USA building codes actually *require* starter bars and stakes—the following explanation of why this practice is not recommend is included.

A common practice has been to insert reinforcing bar (usually 12 mm diameter) vertically into the footings when the footings are being laid. These bars are allowed to protrude about 300 mm into the wall, and are usually called 'starter bars'. The first layer of bales is skewered onto these bars.

To improve the stability of straw bale walls that are not compressed with some sort of vertical tie-down, another common practice has been to use steel, wood or bamboo stakes driven vertically through bales. Most commonly reinforcing bars are used. These are driven vertically down through bales after each layer of bales has been placed. The stakes are typically about 600 mm long and usually two stakes are driven through each bale.

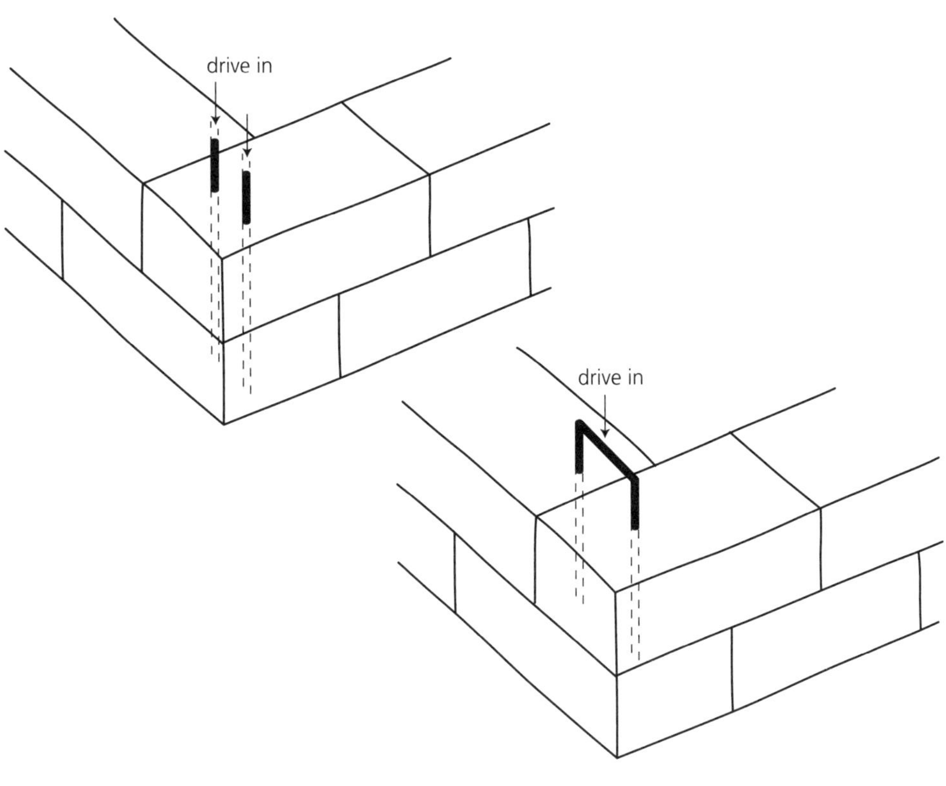


**Fig. 17** Options for pinning bales at corners using 13 mm reinforcing bar or bamboo/wood stakes.

However, compressing the wall using a wire tie-down method firmly locates the wall bottom and makes the whole wall quite rigid. Therefore, starter bars are not necessary to fix the bottom of the wall and, although stakes might improve the stability of the wall during construction, they will contribute very little to the stability of the final wall compared with that achieved by tensioning the tie-down wires.

Starter bars also are dangerous prior to placing the bales and will necessarily penetrate the dampcourse, causing potential sealing problems.

There is an exception, however. Stakes may be used with benefit at corners where bales of adjacent walls form the corner without any other corner structural element. Such a form will not be common with in-fill walls, but will be common with load-bearing walls.

One or two stakes can be driven vertically through the overlapping section of bales forming the adjacent walls after each course of bales is laid. Alternatively or in addition U-shaped bars can be utilised. These will help to keep the walls together and vertical prior to plastering.

# 6

# Stacking, tying and straightening

This chapter describes methods that are in common use, while Chapter 8 describes a new method for tying down straw bale walls that has many advantages over current methods.

In the literature, some descriptions of wall construction methods are quite extreme. The methods presented here are chosen for simplicity, ease of implementation and structural integrity. Most advocated methods satisfy the last of these criteria, but some are severely lacking in the first two.

## Laying the bales

Before laying the bales, enough bales should be prepared for the wall section, as described previously, and a temporary means should be available to constrain the wall to prevent the untied wall from toppling. All the necessary materials, tools and components to complete the tie-down of the wall should be on hand. If there is any risk of rain falling on the wall before the wall is to be plastered, tarpaulins or other weather protection material should be available. There are many references to people allowing rain to fall on unplastered walls and just letting it dry, but if the drying is not rapid some deterioration will occur.

Lay the dampcourse on the raised base, which has been prepared with the chosen method of securing tie-down wires. Place bales in a running bond manner (see Fig. 1, page 5). If the bales vary significantly in length it might be feasible to select bales so that no bales need to be cut to complete alternate layers.

Next comes one of the more time-consuming parts of the quickest part of erecting the wall: making part-bales (see page 24). A part-bale will be required in at least each alternate layer, because of the running bond structure.

After ensuring that all full bales in a layer are firmly pushed together, measure the length of the space remaining. Prepare a part-bale, making the part-bale marginally longer than the space to be filled, except for walls with a free end. In the case of walls with a free end, if the part-bale is smaller than a half-bale the part bale should not be placed at the end, and it should not be very close to a corner.

Frequently during erection of the wall it should be checked for verticality and straightness. A spirit level and/or plumb bob are obvious necessary tools, but one can also observe the wall alignment relative to vertical components of the building frame.

If a whole number of bales does not fill neatly the available space for in-fill walls, or produce the desired height wall for load-bearing walls, straw biscuits may be used to adjust the height. In the load bearing case, trials (similar to the process described on page 76) should be used to deduce the final height after tensioning. It tends to be preferable to insert biscuits just below the top layer of bales, because this preserves a neater top to the wall.

## Installing the top plate and tying down the wall

The wall can tend to tilt during tensioning the wires. There are three significant issues.

- The friction between steel wire and the top plate will cause unequal tension in the wire on opposite sides of the wall if the wire is tensioned on only one side of the wall. This problem is much greater with wooden than with steel top plates, because steel-on-wood friction is much greater than steel-on-steel friction. Wires can be tensioned separately on both sides of the wall, but that is very cumbersome and time consuming. Therefore, a top plate consisting of a ladder made from steel pipe tends to result in fewer wall-tilting problems as the wires are being tensioned. However, later (see Chapter 8) a method that eliminates this problem will be described.
- If the top plate is not very stiff against twisting and bending, the top plate will not transfer the force applied by one wire to adjacent parts of the wall. As a result, the wall is more likely to tilt in various ways as the top plate bends and twists under the uneven tension being applied on opposite sides of the wall. Use of the heavier 25 mm NB pipe in the top plate will help to minimise these effects.
- The third issue relates to the width of the top plate. Relatively narrow top plates will lead to the wire tension being applied closer to the centre of the top of the wall, because the wires cut into the edges of the top bales. However the top plate should still be designed to spread the load across a substantial area of the top of the wall. A suitable compromise is

a 300 mm width for the top plate, thereby spreading the load over ⅔ of the top surface of the wall, and keeping the edge of the plate well away from the edge of the bales.

The first of these issues favours a steel ladder top plate, but the second favours a heavy timber plank, preferably of hardwood. The latter is also advantageous for roof attachment in the case of load-bearing walls, but a relatively heavy steel ladder (say 25 mm NB) may be used with load bearing walls. In the case of in-fill walls one can tolerate some distortion of the top plate, so a relatively lightweight steel ladder can be used in that case.

After all the bales have been laid for a wall section, the top plate can be laid on the top of the wall and tie-down wires installed.

If the wires are to be threaded through tubes in the footing—or in a unit attached to the footing—the wire should go through the tube, over the top plate and the two ends should be overlapped by only a small amount. Join the ends by installing a wire lock. The join should be at a convenient height—about 1.2 metres—for convenient access for using a fence strainer to tension the wires. It is very important that the join of each alternate wire is placed on opposite sides of the wall to minimise the tendency to tilt the wall towards the side being tensioned. The tendency towards unequal tension can be countered to a large extent by pulling the most-tensioned wire out from the wall after each tensioning of the wire, thereby forcing some slippage across the top plate.

After all wires in the section of the wall have been installed and hand tightened—may use pliers, but not wire tensioners—check that the top plate is correctly positioned in the centre of the wall. Check that the wall is exactly vertical, and check that the wall is essentially straight with no significant bulges or misalignments. A large wooden mallet (see Fig. 13, page 21) can be used to persuade the wall to conform to requirements.

Then each wire should be tensioned slightly, and the wall alignment and flatness re-checked and adjusted as necessary. Proceed with the tightening in stages, ensuring that all wires are tightened by about the same amount as each other, and monitoring and adjusting the wall alignment and flatness as tightening progresses. Careful attention to this procedure is very important for the quality of the final product.

## Tie-down procedure for load-bearing walls

The following addresses some of the basic principles relevant to the tie-down of load-bearing walls, *but the actual values of loads mentioned are provided only to indicate the approximate magnitude of loads.* This material should provide a basis for further development of procedures for tie-down of load-bearing walls, but it is not likely to be the final word on the practical application of the methods described.

The tensioning of in-fill walls is not critical. They should be sufficiently tensioned to provide good rigidity and to ensure that the bales will not sag over time. An appropriate choice is a wire tension that produces a force per metre length of wall of approximately one tonne.

However, the case of load-bearing walls is quite different. The roof will not apply an exactly uniform load on all walls, and all parts of walls will not be the same, except in a case such as a simple round building with circularly symmetric roof and completely uniform walls; i.e. no doors or windows—a very unrepresentative and not very useful case. In the common case of an essentially rectangular building with roof trusses, most of the roof load is transferred by the roof trusses to two opposite walls; the other two walls carry very little of the roof load.

Roof loads can vary a great deal in magnitude and distribution depending on the building design and the materials used for the roof. An order-of-magnitude indication is that a metal roof supported by pine wood roof trusses might have a mass of about 40 kg per square metre and a comparable tiled roof might be about twice that mass. Such a metal roof on trusses supported by two walls 6 metres apart would apply a static load of about 120 kg per metre length of wall. Note that these provide only a very rough indication of roof loads, and the figure of 120 kg per metre of wall often will be exceeded.

The following is a basic mathematical treatment of tensioning issues. The reader can skip this section if desired, but a detailed appreciation of this will be of value for the professional builder, particularly for those considering constructing load-bearing walls.

If the final tension in each tie-down wire is T, then the force applied per metre length of wall will be $F = T/s$, where s = space between wires in metres; i.e. $T = Fs/2$. The factor of 2 accounts for there being wires on both sides of the wall. This is an approximation, because the method described above for tensioning the wires will not produce equal tension on both sides of the wall for the reasons already discussed. For an in-fill wall that is tensioned to 1000 kg force per metre—1 tonne, as mentioned above—the wire tension for wires 450 mm apart would be 225 kgf.*

In the case of load-bearing walls, the tensioning wires should load the wall by significantly more than the calculated roof load. Roof loading of walls for a typical modern domestic dwelling with a metal roof is likely be of the order of 120 kg per metre of wall length, as mentioned above. This compares with the wall mass of about 500 kg per metre and a load applied by the tie-down wires of, say, 1000 kg per metre. In fact, the load applied by the tie-down wires easily can be made to exceed considerably 1000 kg/m if that is

* Note: kgf stands for kilograms force. 1 kgf = 9.8 N (newtons).

necessary. If the tie-down wires are 300 mm apart and tensioned to half their breaking strain (for typical 2.5 mm high tensile wire) the load thus applied to the wall will be about 1800 kg/m. Though such a load can be applied quite readily, it is not necessary. However, it does illustrate that the pre-loading applied to the wall before the roof is installed can far exceed the load that will be applied by the roof.

Most roofs will load the walls quite non-uniformly, and particularly if the roof is complex. Before constructing a wall that is to bear the load of a roof, a reasonable estimate of the ultimate roof load distribution on the walls should be made. The aim should be to have a maximum roof load at any point that is well below the pre-loading applied by the tie-down wires.

If the walls were not pre-tensioned before the roof was installed, apart from the fact that the walls would be unstable, the final wall height would be difficult to estimate and, since the roof loading would be non-uniform, the walls would settle to uneven heights. Clearly, not pre-tensioning walls that will be load-bearing is not acceptable.

After all of the wires have been tensioned and before the roof is installed there will be some creep in the straw, which will cause some of the wire tension to be lost, but the wall height should not change measurably during this loss of tension. When the roof is installed any further compression of the walls should be very minor, provided the pre-tensioning was adequate. However, the wires are likely to lose their tension, but should not become slack. If they do become slack, i.e. if the wall compresses significantly when the roof is installed, the probable cause will be that the walls have not been pre-loaded sufficiently.

After the roof is in place, all of the wires should be re-tensioned, but the total load of the roof plus the load applied by the re-tensioned wires should not exceed the pre-tensioning load. The purpose of re-tensioning is only to ensure there is no slack in the tie-down wires, so they will restrain the roof against wind loads. This re-tensioning should be done after adjustments have been made to ensure that the roof is exactly level, for example by using shims between the top plates and the roof structure. Substantial variation in re-tensioning at different locations should be avoided. Some minor final re-levelling might be necessary after re-tensioning, but if it is more than very minor, one should be concerned.

When the roof structure is firmly secured to the top plate, which is firmly tied to the footings with the post-tensioned wires, the whole structure should be adequately secured against wind that might tend to lift the roof. When the wall is plastered, wind-loading forces will be transferred via the wires to the footings, and some of this loading will be transferred to the body of the wall, because the wires will be embedded in the plaster.

## Bulge problems and their solution

During tie-down of walls bulges can occur for several reasons, particularly for high walls. If they do occur they are very difficult to remove without releasing the tension in all wires and starting the tie-down process again after straightening the wall with large mallets and possibly other means of persuasion.

Of course, an essential measure to minimise bulging is to ensure that the wall is straight, flat and vertical before and during tensioning. However, for walls higher than about 2.5 metres it is highly desirable to build and tie down the wall in two stages. Although some commentators and building codes advocate or require that straw bale walls be no higher than about 2.5 metres, much higher walls have been constructed successfully. The following method can be applied effectively to walls at least as high as 4 metres for in-fill walls. However, building load-bearing walls higher than about 3 metres is probably a doubtful proposition until testing enables guidelines to be established for the maximum height of load-bearing walls.

Build the wall to about 2 metres height, or about half the finished wall height, install a top plate and tie down by the usual method. One then has a straight, rigid structure on which to build the rest of the wall. The top plate just installed remains in the wall; it is simply buried. However, since it is to be buried, it is best to use a relatively thin top plate, such as the steel pipe ladder top plate described previously.

Build the second half of the wall in the same manner as the first half, and tie that down with wire ties that reach all the way to the footings. When these are tightened the wires installed on the lower section will become loose, because the lower part of the wall is now under the tension of the new wires plus the weight of the upper section of the wall. Since the wires that were installed in the lower section no longer serve any purpose, they may be removed partially or fully, as convenient, or simply buried in the plaster. At least the wire locks usually would be recovered.

If holes in the footings are used for tie-down wires, this second set of tie-down wires can be threaded through the same holes in the footing—another advantage of the use of such holes.

# 7

# Plastering straw bales

The term 'plaster' in this chapter is used to refer to all mixes of aggregates and binders to form a relatively hard to very hard coating on straw bale walls. 'Aggregates' may be fine to coarse sand and other materials, and usually should be 'sharp', meaning rough and jagged rather than smooth rounded grains. Binders, including clay, lime and Portland cement, bind the aggregates together.

Plaster or 'hard plaster' refers also to similar material that has been used since ancient times to make mouldings and other decorative features on buildings. Modern plastering that involves the fitting of plaster sheeting and pre-made mouldings in modern buildings is not included in the current discussion.

'Stucco' generally refers to a Portland cement-based plaster used on exterior walls, especially in North America, and 'render' is often used interchangeably with the term 'plaster'.

## Plaster types

Earthen plasters and lime plasters are increasingly displacing cement-based plaster for straw bale walls. This has been occurring for various environmental, aesthetic and performance reasons, but some builders will still prefer to use cement, at least for some environments. Their reasons for choosing cement-based plaster sometimes arise from invalid preconceptions. For example, one might choose cement-based plaster to achieve a durable, waterproof external finish, but, for example, despite its generally lower strength, lime plaster has endured for millennia on some external walls.[1] There are also various measures that can be taken in the design and choice of materials that will achieve the necessary durability and waterproofing properties for most applications.

---

[1]See: B Marchese (1980). Non-crystalline $Ca(OH)_2$ in ancient non-hydraulic lime mortars, *Cement and Concrete Research*, 10: 861–64.

## Walls that breathe

It is generally agreed that the plaster should be able to 'breathe'—i.e. it should be permeable to water vapour and air. Water vapour can permeate fairly readily through most lime and earthen plasters, but cement-based plasters with a high cement content tend to have substantially lower breathability.

Excessive build up of moisture in other wall types, such as timber stud walls sheeted with relatively impermeable cladding, also can be a substantial problem that can lead to severe deterioration of the wall. In the case of straw, as with wood, the equilibrium water content tends to be about 12% of the mass when there is easy movement of air and water vapour in and out of the straw, such as with an unplastered straw bale or unclad timber kept away from the weather. This is a surprisingly high figure when one realises that it translates into about 16 litres of water—a good sized bucket full—contained in the straw in each 1 metre length of a typical straw bale wall, and that is desirable! If the water content increases substantially above that figure, microbial growth can occur and the straw will deteriorate. If the water content is high enough (so the straw actually feels wet) the straw will decompose so rapidly that it can become quite warm, or even hot, as one finds in a well-balanced compost heap.

To achieve good water content equilibrium with the atmosphere, and therefore a 'healthy' wall (whether straw bale or wood) it is highly desirable for the plaster to be permeable to water vapour and also desirable that liquid water cannot enter the wall. Throughout the life of a wall one cannot be certain that liquid water will never enter the wall as a result of some deterioration, damage or some unforeseen design error. Apart from taking action to prevent further entry of liquid water, if the problem is detected at all, it is highly desirable that the plaster be water-vapour-permeable enough to enable the water that has entered the wall to evaporate through the plaster at a fairly rapid rate.

Liquid water will also permeate into most plasters to some degree, and some will act like blotting paper and wick liquid water into the wall. If plaster is wet for prolonged periods this could well lead to serious deterioration of the straw. However, the plaster can also act to wick water out of the wall as the outer surface dries. The fact that many plastered straw bale buildings are surviving well, including some of the earliest buildings that are more than 100 years old, provides anecdotal evidence that plaster on straw does serve to protect the straw. However this evidence is not an adequate basis on which to base building codes. More research is desirable in this area.

Based on that anecdotal evidence and the knowledge that liquid water will permeate into most plasters, it seems sensible to design buildings to ensure that the plaster will not be wet for prolonged periods and that the plaster has a fairly high vapour permeability.

One can make some generalisations about the relative moisture barrier properties of the various plaster types, but considerable variation is possible within each plaster type depending on the detailed composition and method of application. For example, steel trowel finishing will tend to compact the surface and produce a smoother surface, which will shed water better and slow the penetration of liquid water. Earthen and lime plasters tend to be quite water-vapour-permeable. Cement-based plasters have quite variable permeability to water (liquid and vapour) depending on their cement content and the grading of the sand.

### Weather resistance

Resistance of plaster to erosion from rain and temperature variations is another important consideration for external plaster. Large roof overhangs can help (and also reduce the degree to which the plaster is wet) but large roof overhangs might conflict with other desired design features, including considerations such as solar passive heating. With moderate roof overhangs, driving rain can still cause problems with plasters that are not resistant to rain.

It is usually desirable to choose external finish plaster that has fairly good weather resistance, even in the case of substantial roof overhangs. The choice of internal plaster tends to be made on the basis of aesthetics and compatibility with use of the building space—e.g. a workshop might be treated quite differently to a dining room, and wet areas need other considerations.

### Important plaster considerations

Important plaster considerations are:

- tendency for plaster to crack, which may allow water to enter the wall, and may be aesthetically unacceptable;
- tendency of plaster to absorb liquid water by capillary action;
- weather resistance, particularly to driving rain and temperature variations;
- permeability of the cured plaster to water vapour;
- compressive and tensile strength of plaster;
- amount of labour and equipment required to prepare and apply the plaster; and
- nature of the finish that can be achieved—e.g. colour, texture, integrity, washability.

It should be apparent from this list of variables that plaster for straw bale walls is a topic that cannot be treated exhaustively within the limited scope of this book, but it is also apparent that there is room for considerable more research in this area. The comments above and following provide only an introduction to the subject.

## Earth-based plaster

Typically earthen plasters consist of a mix of clay and sand and may have added chopped straw or other fibre such as polypropylene fibre. However there are many types of clay. As well, a particular soil sample may contain anything from zero to 100% clay, granular material (e.g. sand or gravel) of various particle sizes, and organic matter. It is rare to find an earth with ideal plaster properties; i.e. one that does not require the clay/sand ratio to be adjusted, contains an appropriate type of clay, has an ideal distribution of particle sizes and does not contain undesirable materials such as a substantial quantity of organic matter.

### The nature of clay

Clays are formed by chemical, biological and weathering actions on rocks, and are composed primarily of aluminosilicates—compounds of aluminium and silicon, with oxygen and hydrogen—but a number of other metallic elements can form part of their composition, such as iron, magnesium, manganese and titanium.

At less than 2 microns (0.002 mm), individual clay particles are not visible to the naked eye. As well as being very small, the atoms in clay particles are arranged in layers, so they have a high aspect ratio; clay particles can have a width to thickness ratio as much as 500:1.[2] Seldom are clay particles regular in shape, and often they have a permanent electrical charge, depending on their composition. The very small particle size and the irregular and layered structure of the particles cause clays to have extremely large surface areas, even as much as 1000 times the surface area of sand. The combined surface area of all the clay particles in one teaspoon of clay can be as large as a rugby field!

Knowledge of these physical characteristics helps us to understand the macroscopic properties of clay. Clays attract water—they are hydrophilic—because of the net electrical charge on the clay particles. Some clays can absorb five time their own volume of water, but some other materials of related chemical composition and composed of equally fine particles, such as talc, which does not have a layered structure and carries no charge, will repel water—talc is hydrophobic.

This affinity of clay for water leads to the properties of clay of most interest to us:

- swelling and shrinking, with the resultant tendency to crack as it dries;
- plasticity of damp clay;
- cohesion (and also complete loss of cohesion in the case of some clays that continue to absorb water until they liquefy); and
- absorption of water by clay that has been dried, such as a finished and dried plastered wall.

[2] Smectite particles can be 0.001 microns thick by 0.5 microns wide.

The result of swelling and shrinking as clay absorbs water and dries is graphically illustrated by such unwanted effects as cracking of buildings in areas where there are clay-rich soils and extended dry periods, such as Adelaide, South Australia. The curators of cricket pitches take advantage of this property to achieve a hard, crazed surface to the pitch.

However, absorption of water by clay can actually prevent water from passing through the clay. This effect is utilised in the use of bentonite clays to seal earth dam walls by swelling and closing pores between other soil particles.

Some clays will swell much more than others. For example smectites tend to swell much more than kaolinite, and the swelling in smectites is highly dependent on the amount of sodium or calcium in the clay—in that case sodium enables much greater swelling.

The main property of clay of interest for plastering straw bale walls is its expansion/contraction characteristics as water content changes, and the tendency of the dry plaster to absorb liquid water.

## The effect of sand and silt in an earthen plaster mix

Sand comes in a variety of specifications—coarse, coarse washed, sharp, plasterers', brickies', various grain sizes and clay contents, and various colours from nearly white to reds, browns, etc. Sand is defined[3] to have grain sizes in the range 20 microns to 2 mm. Above 2 mm we enter the gravel category. However, material sold as coarse sand often contains grains up to about 4 mm. Silt has grain sizes in the range 2 to 20 microns. Clay particles are less than 2 microns.

Sand and silt give structural strength and hardness to the plaster. The clay is the cement or binding agent that holds the sand and silt together. If the non-clay component is entirely composed of large sand particles, say 1 to 2 mm, when these particles pack together there will be a lot of space between them with only clay to fill the space. In thin films between sand particles the clay has a strong cementing effect, but the clay that fills the larger voids exhibits much less strength, though this depends on the type of clay. If the sand is 'sharp', it will be more effective in producing a strong, hard plaster. Sharp sand is composed of irregular as distinct from rounded particles.

The smaller sand particles and the silt fill the voids between larger sand gains, so the clay then fills only very thin spaces between these closely packed sand and silt particles. It is easy to appreciate that this close packed structure of particles with a wide range of sizes, cemented together with clay particles that are not visible to the naked eye, is likely to be strong and hard.

The ideal plaster composition will have a fairly uniform distribution of sand and silt particles from 2 microns to about 1 mm. Such sand is said to be 'well graded'. Such clay-silt-sand mixes occur naturally in some locations, such

[3]The definition is slightly different in the USA, and is expressed in fractions of an inch.

as in parts of south central New South Wales, Australia, where sand and silt has been deposited by wind that has carried the material from western arid regions and clay has been incorporated by the action of water. These processes occur over hundreds and thousands of years.

## Selecting earthen plaster materials

An earthen mix consisting of about 25% to 30% clay and the rest being sand and silt, but with very little or no organic material, is likely to be a usable plaster mix. A mix of these proportions will feel much more like sand than clay. However, the type of clay and the distribution of particle sizes in the silt and sand fraction will affect the properties of the plaster.

A common need is to adjust the clay-sand ratio. Usually it is easier to add more sand than to add more clay, because sand is easier to dig, to handle and to buy.

If earth for plaster is to be sourced in the field, apart from considerations such as whether the collection area will damaged, erosion caused, etc., first simple field tests should be done to determine roughly whether the soil is suitable.

Soil is nearly always layered. A cut into the soil to expose a vertical section will reveal the layers. Weather erosion might have already exposed a vertical section where little effort is required to cut to a clean face. Colour and texture changes with depth should be readily visible.

The topmost layer usually will be topsoil, which will not be suitable for plaster, because it will contain excessive quantities of organic matter. If the topsoil is deeper than too deep to dig easily, the site is not suitable and another site should be sought.

If it is easy to dig beyond the topsoil, then the following tests should be done.

### Appearance and initial feel:

- If it contains rocks and grains that are clearly more than 2 mm probably it should be sieved, so usually another source should be sought.

### Colour:

- Dark grey to black will suggest substantial organic matter content. Such soil should be avoided, because the organic matter may decompose with time, it generally weakens the final product, and adds no benefit.
- Light colour soil is probably low in organic material and low in iron.
- Red and yellow soil that has not been darkened with organic material will have some colouring mineral such as iron compounds.

As long as the colour does not indicate substantial organic content, and has a colour that is aesthetically acceptable, one can move to the next test.

## Feel:

- Rub it gently between the thumb and fingers to determine whether it feels sandy, smooth, silky, spongy, plastic, or a combination of these. The earth should be slightly damp for this test. If there is no sandy feel to it, there is not enough sand. If it feels like sand with very little coherence when damp, it will have too little clay. If it is smooth and plastic, it is probably mostly clay, but if it also has some silkiness to the feel it probably contains a substantial amount of silt. If the soil feels greasy, that is an indication of organic matter, which is undesirable. The ideal material will feel sandy, but also have some smooth, silky and plastic components to the feel. It will be fairly coherent when damp. Refer to the field texture grades table for a more detailed description.

From a vertical face, take handful-sized samples of the soil from various measured or marked depths in the region where tests have indicated that the soil might be suitable. Dampen the sample uniformly, but do not make it wet. If it is anywhere close to the desired composition, it will be possible to mould it by hand without it collapsing. If it collapses easily it is either too dry or too sandy. Kneed the sample into a ball roughly 30 mm diameter, and attempt to extrude this as a 5 mm thick horizontal ribbon by working it between thumb and forefinger. The length of extruded ribbon that can be achieved will give a rough indication of the clay content of the sample. It should be possible to form a ribbon of about 30 to 40 mm length if the earth is suitable.

These tests may be followed with a settling test. Take a straight-sided glass jar, such as a food-preserving jar, about three-quarters fill it with the chosen soil, and then nearly fill it with clean water. Shake the jar until the water and soil have thoroughly mixed. Then let the jar stand undisturbed for several hours.

The larger sand particles will settle at the fastest rate and the smallest clay particles might never settle. After the settling is complete, there will be clearly visible layers of the coarsest material at the bottom, and finer and finer material in the higher layers. If the water becomes perfectly clear there are no clay particles at the very fine end, i.e. less than about 0.01 microns; that is of no concern either way. These extremely fine particles tend to remain in suspension as 'colloidal' particles, because they are light enough that the motion of water molecules due their thermal energy is sufficient to keep the particles moving randomly in the water. This is only of academic interest in the current context.

Ideally there will be a layer of sand that merges into silt, with a combined depth of about 70% of the total solids, the rest (about 30%) being clay. If in doubt as to whether the top layer is clay or silt, carefully extract a sample, allow it to dry until it is only damp, and then feel it. If it is clay, it will have a smooth feel without any of the silky feel of the fine silt.

Table: Field texture grades of soil

| Field texture grade | Behaviour of moist bolus | Approximate clay content |
|---|---|---|
| Sand | Coherence nil to very slight, cannot be moulded; single sand grains adhere to fingers | Always less than 10% and commonly less than 5% |
| Loamy sand | Slight coherence; can be sheared between thumb and forefinger to give minimal ribbon of about 5 mm | 5% to 10% |
| Clayey sand | Slight coherence; sticky when wet, many sand grains stick to finger; will form minimal ribbon of 5–15 mm; discolours fingers with clay stain | 5% to 10% |
| Sandy loam | Bolus just coherent but very sandy to touch: will form ribbon of 15–25 mm; dominant sand grains are of medium size and are readily visible | 10% to 15% |
| Light sandy | Bolus strongly coherent but sandy to touch; sand grains dominantly medium size and easily visible; will form ribbon of about 20–25 mm | 15% to 20% |
| Loam | Bolus coherent and rather spongy; smooth feel when manipulated but with no obvious 'sandiness' or 'silkiness'; may be somewhat greasy to the touch if much organic matter present; will form ribbon of about 25 mm | About 25% |
| Silt loam | Coherent bolus; very smooth to silky when manipulated; will form ribbon of about 25 mm | About 25% and with silt 25% or more |
| Sandy clay loam | Strongly coherent bolus, sandy to touch; medium size sand grains visible in finer matrix; will form ribbon of 25–40 mm | 20% to 30% |
| Clay loam | Coherent plastic bolus, smooth to manipulate; will form ribbon of 40–50 mm | 30% to 35% |
| Silty clay loam | Coherent smooth bolus, plastic and silky to the touch; will form ribbon of 40–50 mm | 30% to 35% and with silt 25% or more |
| Sandy clay | Plastic bolus; fine to medium sand grains can be seen felt or heard in clayey matrix; will form ribbon of 50–75 mm | 35% to 40% |
| Silty clay | Plastic bolus; smooth and silky to manipulate; will form ribbon of 50–75 mm | 35% to 40% and with silt 25% or more |
| Light clay | Plastic bolus; smooth to touch, slight resistance to shearing between thumb and forefinger; will form ribbon of 50–75 mm | 35% to 40% |
| Light medium | Plastic bolus; smooth to touch; slightly greater resistance to ribboning shear than light clay; will form ribbon of about 75 mm | 40% to 45% |
| Medium clay | Smooth plastic bolus; handles like plasticine and can be moulded into rods without fracture; has some resistance to ribboning shear; will form ribbon of 75 mm or more | 45% to 55% |
| Heavy clay | Smooth plastic bolus; handles like stiff plasticine; can be moulded into rods without fracture; has firm resistance to ribboning shear; will form ribbon of 75 mm or more | 50% or more |

Source: Natalie Hunt and Bob Gilkes (1992). *Farm Monitoring Handbook*, University of Western Australia.

## Shrinkage properties of the plaster mix

As wet soil that contains clay loses water it will shrink. Initially that shrinkage will equal the volume of water lost, but as more water is lost it will reach a point where the shrinkage is not as great as the volume of water lost. Generally the less clay content the more water can be lost before this transition occurs, but it also depends on the type of clay. A normal fresh plaster mix already will have water content about or below this point, and therefore this initial shrinkage proportional to the water loss is not of great significance for plaster.

Pure clay will shrink more, usually much more, than clay-sand mix. The thinner the films of clay within the clay–sand–silt mix, the less will be the shrinkage. As the mix is adjusted to achieve closer packing of the sand and silt particles, there will be left fewer places where these particles are not in intimate contact, thereby preventing further shrinkage due to the clay component shrinking as water is lost from the region between clay particles.

Therefore, although the clay in a plaster mix might have large shrinkage characteristics, when combined with the sand and silt the shrinkage can be minimal.

### Shrinkage testing

Soil science texts describe simple test methods, typically using half-cylinders of copper with their ends closed with soldered copper sheet. An alternative is apparatus consisting of plastic drainpipe that has been halved lengthwise and can be assembled in a wooden jig that clamps flat metal sheet (e.g. aluminium) or glass on the ends. This design uses readily available components; it can be made easily to have the ends accurately parallel so the initial length of the sample can be measured accurately; and the sample can be removed very easily after it dries by dismantling the jig. Before use, the components should be greased, e.g. with Vaseline, to aid extraction of the dried sample. Plastic pipe dimensions of about 50 mm diameter and about 100 mm long are suitable.

The examples of shrinkage tests in Fig. 19 dramatically indicate the major shrinkage difference between pure clay and clay/sand mix.

### Trial use of plaster

The final stage before using the plaster should be to test the plaster by applying it to a small section of straw bale wall in a manner and an environment similar to the planned application. It is important to understand that 'environment' includes factors such as the temperature, humidity, presence or absence of wind, and aspect of the wall; e.g. is it an internal or external wall, and will the sun fall directly onto the wall during the early stages of the drying/curing of the plaster? All of these factors will affect the rate of drying of the plaster and therefore potentially affect the final result, such as whether cracks or crazing will occur.

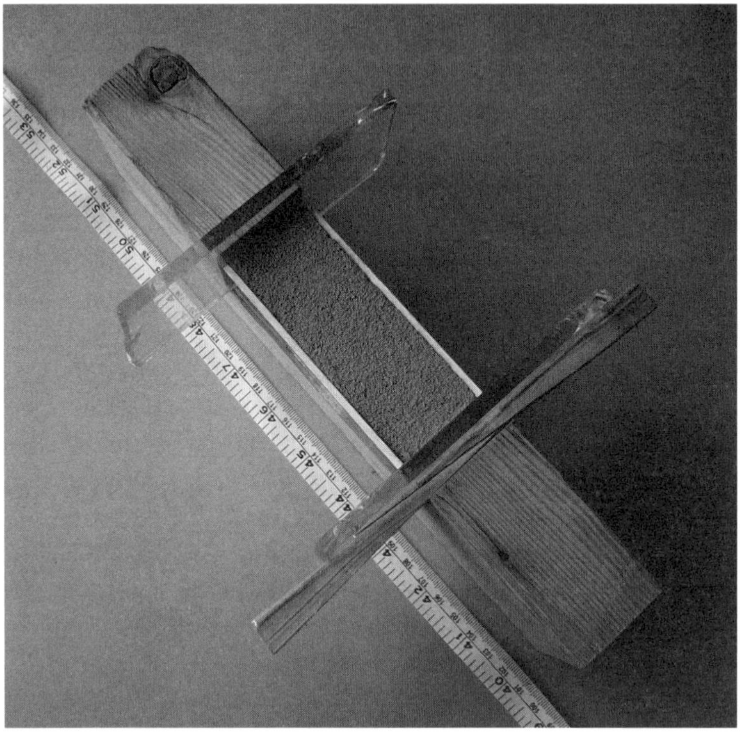

**Fig. 18** This simple apparatus is useful for testing the shrinkage of plaster.

**Fig. 19** A shrinkage test demonstrates the dramatic reduction in shrinkage due to the presence of sand. The lower sample is approximately 70% sand, 30% clay, and the upper sample has had all of the sand component removed.

Fresh plaster on an external wall subject to sun and hot wind is likely to behave in a vastly different way to fresh plaster on an internal wall in cold damp winter conditions. The latter might not dry for weeks, and could grow mould; seeds left in the straw might even germinate; and straw immediately adjacent to the plaster might begin to rot, though the low temperature will help to inhibit rotting of the straw. The former might dry and crack in hours.

If three layers of plastering are to be done, then the full procedure of three coats should be tested.

This trial run might take some days or more to complete and it might be necessary to run different trials on internal and external walls, and perhaps on north and south facing walls. Therefore, it could well be desirable to undertake such tests prior to erecting any walls.

These tests are particularly desirable in the case of earthen plaster, because of the great variability of materials that can be utilised. The detailed composition of lime-based and cement-based plasters tends to be much more accurately known, but similar tests are desirable if data on tests already undertaken for the chosen mixes are not available.

Earthen plasters that do not have adequate weather resistance may be finished with a lime/sand plaster, possibly with the inclusion of additives such as linseed oil, or by hydrophobic treatment (see page 57).

## Lime-based plaster

In the context of plastering straw bale walls, lime is hydrated calcium hydroxide ($Ca(OH)_2$) plus impurities and, sometimes, additives. However, lime can refer to 'agricultural lime', which is calcium carbonate ($CaCO_3$), which is not used in plastering, and to 'quicklime', which is calcium oxide ($CaO$).

The use of the word 'hydrated' for $Ca(OH)_2$ can be misleading, because people with some knowledge of chemistry will recall that a hydrated molecule usually refers to a one to which a certain number of water molecules are attached, the material still existing in a dry form. Use of the term 'hydrated' here refers to the fact that hydrated lime is formed by the reaction of quicklime with water to form molecules ($Ca(OH)_2$) that have no attached water molecules.

Lime exists naturally as calcium carbonate in limestone and chalk, which are common minerals.

Quicklime (also called burnt lime, unslaked lime or caustic lime) can be manufactured from limestone and typically contains 80% to 90% calcium oxide, the rest being impurities carried over from the mined limestone. These impurities commonly comprise oxides of silicon, magnesium, aluminium and iron. Heating limestone to about 500°C, which causes it to decompose into calcium oxide and carbon dioxide, makes quicklime.

Quicklime can be converted into calcium hydroxide by reacting it with water, which produces a violent reaction that is accompanied by the generation of a large amount of heat and resultant steam; that process is referred to as 'slaking' lime. When CaO (quicklime) reacts with excess water, 1 kg of CaO will combine with 0.32 kg of the water to produce 1.32 kg of $Ca(OH)_2$ (hydrated lime), which will be suspended in any excess water. The amount of excess water will determine the thickness of the lime putty so produced. For example, one commercial lime putty is made from a mix of 38% CaO and 62% water (by weight), which produces a lime putty that is 50.2% solids ($Ca(OH)_2$) and 49.8% excess water. Quicklime is a hazardous material and should be handled only by appropriately trained and experienced people.

Hydrated lime can be in the form of a white powder (builders' lime, brickies' lime, industrial hydrated lime, slaked lime) or as thick white slurry (lime putty, which typically contains about 50% solids). The resultant lime putty can be dried to produce hydrated lime, which may have additives incorporated, such as air-entraining agents, which are included in some builders' lime products, but not all of them. The usual process uses an air-swept mill to directly lift the hydrated lime from the precipitate formed by slaking lime, thereby directly producing a fine powder without the need for any grinding process. The hydrated lime can have some colouration due to impurities such as iron oxide, but such colouring usually is not very significant.

Mixing hydrated lime and water is not accompanied by chemical reaction or by significant dissolution of the hydrated lime, which has very low solubility in water (about 20 grams/litre), but rather results in a suspension of the lime in water with the same chemical composition as lime putty, but not the same flow properties, or rheology. However, the rheology of lime putty prepared by slaking lime depends on the characteristics of the original limestone and the details of its burning and slaking. The ageing of lime putty, meaning retention under excess water for extended periods from months to many years, may homogenise the differences among various slaked limes.[4]

The rheology of a lime putty changes as it ages, because the hydrated lime particles very gradually break down into smaller particles and the shapes of the particles change due to changes in crystal structure towards a more plate-like form.[4] One might expect that such changes should occur in lime putty whether it is produced by slaking lime or by making the putty from hydrated lime powder. However, it appears that the latter begins with a handicap of larger particles[4], which perhaps are also different in structure. Anecdotal evidence is that it takes a long time, perhaps a very long time, to catch up as far as

[4]See: O Cazalla, C Rodríguez-Navarro, E Sebastián, G Cultrone & M J de la Torre (2000). Aging of lime putty: effects on traditional lime mortars carbonation, *Journal of the American Ceramic Society*, 83: 1070–1076.

rheology is concerned. However, currently the research does not appear to match the folklore of traditional craftspeople; further research would be highly desirable to clarify this issue.

After Portland cement-based plaster became so popular since the 19th century, there has been limited scientific research attention given to lime putty and lime plaster. However, the necessity to use lime plaster in the restoration of historic buildings for practical, as well as aesthetic, compatibility reasons has led to some renewed research interest.

## How does lime plaster set or cure?

Lime-based plaster (lime/sand/silt) sets by drying and undergoes some conversion to calcium carbonate very close to the surface during its drying.

In the literature, particularly on the Internet, widely divergent statements can be found about the way that lime in lime-based plaster changes over time from calcium hydroxide to calcium carbonate due to its interaction with atmospheric carbon dioxide, a process called carbonation. These range from claims that lime plaster sets by carbonation[5] to the scientific observation that lime in mortar from Roman times (about 2000 years old) still has not fully converted to calcium carbonate.[6]

The fact[7] is that the setting of lime plaster occurs almost entirely by drying—simply the evaporation of water. A very thin surface layer will carbonise quite quickly (typically in a matter of days) but that layer then acts to inhibit deeper carbonisation and also the evaporation of free water retards the carbonisation, because the reaction with the atmospheric carbon dioxide depends on the presence of water. Cazalla et al.[7] have examined the ageing and carbonation of various lime putties and provided a detailed analysis of their properties and changes in their crystalline structure, and measured the rates of carbonation of a range of samples.

The conversion of calcium hydroxide to calcium carbonate occurs through two stages. First the carbon dioxide in the air is dissolved in water to form carbonic acid ($H_2CO_3$), a very weak acid, which in turn reacts with the calcium hydroxide to form calcium carbonate. The rate that this reaction occurs in plaster after its initial drying is slow, because the concentration of both water and carbon dioxide is low. The rate depends on the humidity (it is faster when there is high relative humidity), the exposure of the plaster to liquid water (e.g. via rain), the atmospheric concentration of carbon dioxide (it

[5]See: http://www.buildingconservation.com/articles/limedefn/limedefn.htm, an article about lime putty.

[6]R Malinowski (1979). Concretes and mortars in ancient aqueducts, *Concrete International*, 1: 66–76; B Marchese (1980). Non-crystalline $Ca(OH)_2$ in ancient non-hydraulic lime mortars, *Cement and Concrete Research*, 10: 861–64.

[7]See: O Cazalla et al. (2000). *Journal of the American Ceramic Society,* 83: 1070–1076.

might, for example, be higher in some industrial areas) and the microstructure of the plaster. Carbonation can be accelerated greatly by putting concentrated carbon dioxide gas in direct contact with wet plaster,[8] but this is only of academic interest since it is not likely to be practicable to do this during normal building construction processes.

As a result it can take many decades for the lime binder in lime plaster to convert substantially into calcium carbonate. However, if the plaster has been adequately protected during the period of substantial carbonising, for example by application of limewash (lime mixed with materials such as tallow and linseed oil) every couple of years, the resultant calcium carbonate has very low solubility in water and therefore little tendency to weather. Calcium carbonate dissolves at the rate of only 1.5 milligrams per litre of water, which is 13 000 times less soluble than calcium hydroxide, but it is complicated by the fact that calcium carbonate in turn reacts with carbonic acid to form calcium bicarbonate ($Ca(HCO_3)_2$), which is much more soluble in water.

Traditionally lime-based plaster was periodically coated with limewash, and such finishes have endured for as long as 2000 years. The fact that the durability of lime plaster has long been known has not limited the almost universal use of Portland cement-based plasters during the last century, because the cement product achieves durability without the need to treat it repeatedly by any equivalent to lime-washing.

Cazalla et al.[9] came to useful conclusions for the conservation of historical buildings, but their conclusions are also relevant to other plastering, such as plastering straw bale walls. Plaster made from very long term aged lime putty (14 years) carbonised considerably faster than plaster made with 1-year aged lime putty, but the latter showed similar carbonation characteristics as plaster made with commercial hydrated lime putty (non-aged).

However, 'if aged lime putty is not available, commercial lime powder can be used if a putty is prepared before mixing with the aggregate'.[9] The basis of this conclusion is their observation that plaster made with lime putty prepared from hydrated lime powder had much lower porosity than plaster prepared by mixing the dry powder with the sand rather than first making a lime putty. They did not age the lime putty prepared from hydrated lime powder. Currently only anecdotal evidence is available on whether some ageing of this type of putty is beneficial; some builders claim that it should be aged for several weeks, but that appears to be a very short period to achieve any significant change on the evidence of the above quoted experiments.

[8]See: DR Moorehead (1986). Cementation by the carbonation of hydrated lime, *Cement and Concrete Research*, 16: 700–708.

[9]See: O Cazalla et al. (2000). *Journal of the American Ceramic Society*, 83: 1070–1076.

## Aggregate to binder ratio

The optimum aggregate (sand and silt) to binder ratio is relevant to all plasters, whatever the binder. Experience has shown that the binder component in the case of earthen plasters (i.e. the clay component) should be about 30%. Cazalla et al.[10] recommended 1:4 when using aged lime putty and 1:3 when using fresh lime putty made from powdered lime—i.e. 25% and 33% binder respectively.

Due to the different degree of packing of different aggregates, depending on the particle shapes and size distribution, one can expect the optimum amount of binder to vary from one aggregate to another. One proposal for determining the optimum for a particular aggregate is to measure the amount of water that the dry aggregate can accommodate. This can be measured easily by a surface area and void test in which equal volumes of dry aggregate and clean water are measured (It might be necessary to heat the aggregate and then allow it to cool to ensure that it is dry). The measured water is then added to the aggregate until the aggregate is just filed with water and no excess lies on top of the aggregate. The volume of added water provides a measure of the void space in the aggregate that should be filled with binder. Less binder than this will lead to places where the aggregate is not 'glued' together; more binder than this will lead to the creation of volumes with excess binder, which are likely to weaken the structure of the plaster and also cause a tendency to crack. Of course, this assumes that the binder will flow into the voids as will water, which is not entirely correct, because the binder particle sizes are much larger than water molecules. This test should therefore provide an upper limit to the amount of binder that should be used.

For the practical purposes of a lime–sand plaster for use on straw bale walls, there are a wide variety of opinions and there is a lot of room for research, but at this stage the following appear to be reasonable conclusions.

- Lime plaster is generally more resistant to weathering than are earthen plasters, and lime plaster is relatively easily repaired if weathering or cracking does occur. To some extent lime plaster is even self-healing for small cracks, because dissolved lime tends to re-deposit in small cracks. One can retain a small quantity of the plaster mix under water in sealed containers for the life of the building, to be used to repair almost invisibly any cracks or damage. By retaining a small quantity of the original plaster mix one can ensure a good colour and texture match.
- Hydrated lime powder sold as builders' lime is suitable, but it should be mixed with water to make a putty (about 1:1 lime powder:water by weight) before adding it to the aggregate. Probably one should choose a

[10]See: O Cazalla et al. (2000). *Journal of the American Ceramic Society,* 83: 1070–1076.

hydrated lime powder that does not have air-entraining additives; the manufacturer should be consulted to ascertain this, because it is not necessarily stated on the packaging. The risk is that air-entraining agents may cause the cured plaster to be more porous. Hydrated lime powder can vary significantly from one manufacturer to another, from one batch to another, and sometimes even from one bag to another. For many applications this will not be significant, but if small colour changes are of concern this could be a problem.

- Lime putty is a good product, but, unless it is aged for several years it appears that the resultant plaster will not have a great advantage over one using a putty made from hydrated lime powder, thought it appears that the flow properties of the two mixes may be quite different. Commercial lime putty is much more expensive than hydrated lime powder (as much as five times) and there are very few suppliers. Lime putty can be made by slaking quicklime, but that process is too hazardous to be used by the average builder—whether professional or owner-builder.
- Lime putty is generally less brittle than Portland cement-based plaster and it has the consequent property that it can creep and flow to accomodate gradual movement of structures over time.

## Portland cement-based plaster

Portland cement is produced by mixing calcium carbonate, silica, alumina, iron oxide and gypsum (calcium sulfate), which is fired and ground to form the cement powder. This is a highly energy–hungry process, which is a common environmentally-based reason for advocating minimal use of Portland cement.

However, Portland cement is a very versatile material, with many options for mixes and many additives designed to produce concrete and plaster with a vast range of properties. Although the various mixes and additives can tailor cement products to achieve a wide variety of properties, such tailoring is inherently more complex than a simple lime-sand-water mix. As a result, cracking of cement-based plaster on straw bale walls is very common.

As much as one might stress that such cracking can be avoided by using the correct mixes and processes, cement-based plasters remain less attractive, because ideal results seldom will be achieved, except by experts. But it remains that absolute assertions that cement-based plasters are not suitable for use with straw bale walls probably are unwarranted generalisations.

For those who do desire to use cement-based plaster, the following considerations may be of value.

- Cement-based plaster can range from quite porous to liquid water, to almost waterproof; towards the porous end of the range, there can be a strong tendency for water to be wicked up or into the plaster by capillary action.

- Since cement-based plaster tends to by more brittle, it is wise to use expansion joints on long sections of wall to minimise the tendency to crack.
- If cracks occur in the plaster, there is not a simple way to repair them, contrary to the situation with lime and earthen plaster.

### Hydrophobic treatment of plaster

The absorption and wicking of water into plaster can be countered by the application of penetrating hydrophobic (water repelling) treatment after the plaster has dried (for example, 'organo-silanes' such as iso butyl triethyloxysilane). This is at least the case for cement-based plaster for which there are available commercial products that are highly penetrating and designed to have insignificant effect on the surface appearance and do not seal the surface. Pores in the plaster remain, but they are all coated with a chemical that repels water. This provides a long-term solution to water permeability of cement-based plaster, because it is applied as a solution that is absorbed deeply into the plaster, rather than being a surface treatment. The same treatment should be suitable for lime-based plaster and perhaps could be applied to earthen plasters, but perhaps this conjecture should be confirmed by tests.

### Drywall finishes for straw bale walls

Sheeting internal and/or external walls with appropriate standard drywall sheeting products is a possibility that has not been significantly researched, but further consideration of this option is deferred to the discussion of a new straw bale wall construction system, because that system can include special features to cater for this option.

## Provisions for attaching loads to plastered walls

There appears to be no simple way to deal with attaching loads to plastered straw bale walls as they have been built to date. Pre-determined hanging points can be installed at the time of construction by measures such as driving into the wall at specific locations, substantial wedges of timber, usually at least to half the wall thickness. The blunt end of these, roughly flush with the straw face, may be plastered over, but if they are to be used for hanging their exact location must be recorded, or marked in some way.

Alternatively an anchoring device of timber or steel can be inserted into the un-plastered wall using plates on opposite faces and bolting these together through the wall. Generally these would, of course, be capable of carrying substantially heavier loads.

Reliable load tests for particular configurations of such anchor systems appear not to be available. Until there is good engineering data on loads that may be applied safely to such anchor points, very conservative loading should be the rule.

Better options for attaching loads to walls using intermediate structure units are discussed in the next chapter.

# 8

# A new straw bale wall construction system

This chapter describes a new straw bale wall construction system that simply and reproducibly produces straight, flat walls that can incorporate features and facilities that are taken for granted in conventional building systems—features such as full-height in-fill walls that do not require special filling at the top, built-in facility for attaching hanging points to the wall, integral skirting board systems and the option of modular construction methods. The principles of the method and some practical details are described, but the system is in the process of development, so there is room for ingenious input to the system design.

The system incorporates a new tensioning system that replaces the use of fence-strainers for tensioning the hold-down/compression wires and eliminates uneven tensioning on opposite sides of the wall. It also includes an option of fabricating wall modules on-site, but as horizontal modules that are then swung into the vertical wall position after assembly. The last option is treated somewhat separately, because it is a more radical change from current methods.

The system has substantial advantages over current methods.

- Wires are simultaneously tensioned on both sides of the wall so that both sides are compressed by exactly the same amount. As a result, there is no wire slippage across the top plate and therefore the tendency for the wall to tilt, bulge or otherwise distort during tensioning the wires is almost completely eliminated, even for thin walls using new smaller bales, which are discussed later.
- Each section of wall can be tensioned uniformly very easily, accurately and quickly.
- Lighter weight top plates may be used.
- New base and intermediate structure units can be used to streamline procedures, provide hanging locations for shelving, etc. and make it more feasible to use drywall cladding as an alternative to plaster.

## The wire tensioning system

The new strainer for tensioning the wires consists of tubes within the base unit, strainer rods, and wrenches.

Steel tubes are incorporated into the base of the wall and the tensioning wires are secured with wire locks attached to a looped and twisted wire attached to the base unit. Re-usable tensioning rods, which are inserted into the tubes in the base, are rotated using a wrench to tension the wires. Multiple tensioning rods are required.

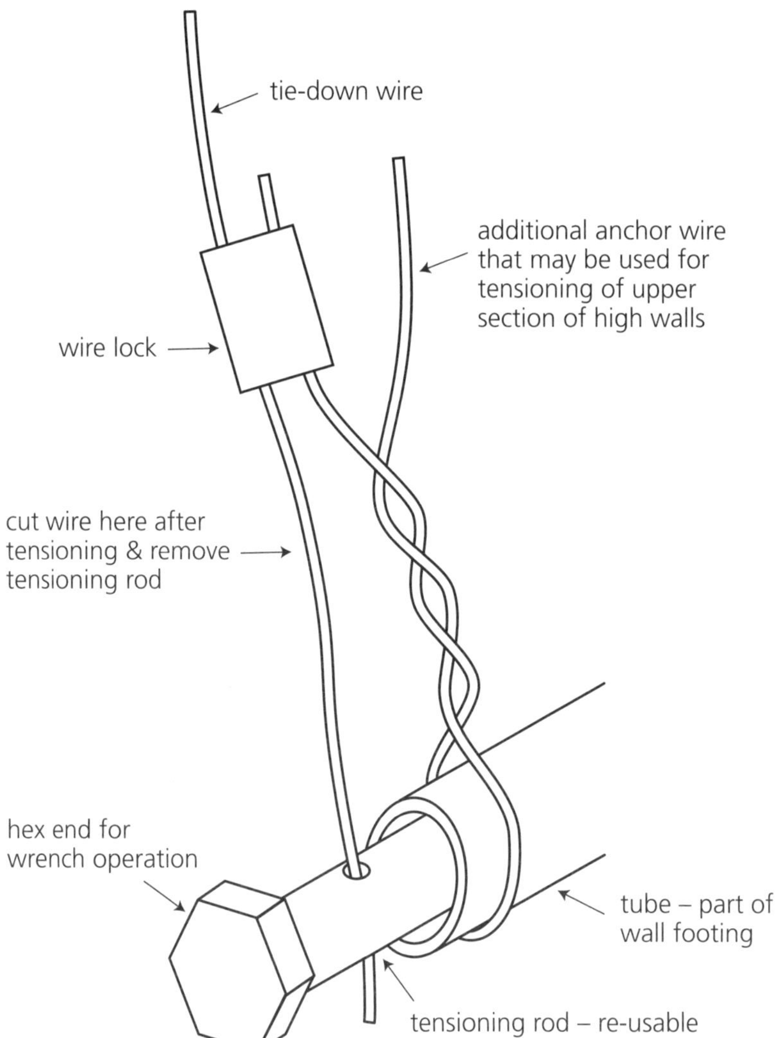


**Fig. 20** Wire tensioning system. The wires are tensioned by rotating the tensioning rod, and the wire is retained by the wire lock and twisted anchor wire. The tensioning rod may be removed after tensioning is completed.

It is advantageous to use a tension wrench to rotate the tension rods, particularly for load-bearing walls, where it is more important to know the load being applied to the wall by the hold-down wires.

Steel tensioning tubes usually would be incorporated into the wall base unit, but they can be incorporated into a raised section of the footing. A tube spacing of 450 mm is suitable, but a closer spacing of say 300 mm is desirable, because the additional wires provide greater control, enable lower wire tension to be used, and the extra effort in tensioning more wires is minimal, because of the ease with which the tensioning can be done.

After a wall section is erected, tensioning rods are inserted into all of the tensioning tubes in that section. Wires are installed over the top plate, through the wire locks on each side of the wall and through the slots/holes in the ends of the tensioning rods. Each of the wires should be pulled firmly through the wire locks with pliers to remove most of the slack in the wires.

It is then a simple matter to tension the wires by turning each tensioning rod with a wrench. The wires should not be tensioned until the wall is straightened and set exactly vertical. All bulges should be removed, but the final trimming of the wall flatness is best done after a small tension has been applied to all of the wires.

Builders who are used to the old method of tying down walls with fence strainers will be amazed at the precision, smoothness and speed of this new method. Using one wrench one can quickly and easily move back and forth along the wall section, tightening each rod only a small amount at a time, which is dramatically quicker, easier and more accurate that using fence strainers. Of course, more than one wrench may be used if more than one operator is available. The ease with which the wrench may be moved from one location to another minimises any inclination of the operator to apply a large amount of tension during each operation of the wrench, and also a closer spacing of wires does not cause significant additional effort. Thereby the wall is compressed very uniformly, both on opposite faces and along the wall length.

The tensioning rods should be withdrawn only when one is satisfied that the tensioning of the wall section has been completed, because to remove a tensioning rod one must first cut the wires just below the wire locks, which effectively precludes re-tensioning unless one removes the wire and inserts a new wire.

When constructing walls that are to be load-bearing, the tensioning rods should not be removed until the roof has been fully installed; that is, when the full roof load is bearing on the walls. Therefore, many more tensioning rods are necessary for a structure with load-bearing walls—enough for the whole building.

## The wall components

The full set of system components are:

1. Straw bales, which may be the conventional 450 × 350 × 900 mm bales, or smaller bales (see pages 64–68).
2. A wall base unit that incorporates tubes at regular intervals (typically 300 to 450 mm) and incorporates means of holding wire locks. This unit must be fixed to the footings in a manner that prevents its movement or distortion during construction or service. This base unit may also incorporate features that enable the easy installation of a modular skirting board system.
3. Intermediate structure units that may be incorporated between particular layers of bales, to be used for fixing a range of things to the finished wall—such as artwork, shelving, and drywall panelling in lieu of plastering—and be used for high walls to provide greater control during erection.
4. A top plate of standard configuration.

### Wall base unit

The system requires tubes to accommodate tensioning rods to be fixed in the base at the desired wire spacing, typically 300 to 450 mm, and wire locks to be fixed with respect to the base unit at each tube location.

A simple form of base unit uses steel tensioning tubes in raised sections of footings that are poured integrally with the rest of the footing. Another alternative of tubes incorporated into a steel ladder base unit is illustrated in the next chapter. Any configuration that satisfies the above basic requirements may be utilised.

The wire lock holders are simple wire loops with twisted ends, which may be used for either one or two wire locks. The twist in these wire lock holders should be tight (less than 20 mm pitch for each wire), symmetrical (both wires bent into the twist, rather than one wire twisted around the other straight wire) and at least three twists of each wire.

These requirements apply, because when only one wire lock is attached to the holder the twist must remain intact when tension is applied. However, it should be noted that the full tension force would not be applied to the wire lock holder, because the full tension force is applied via the tensioning rod to the wire that passes through the wire lock, so no tension is applied to the wire lock holder until the torque ceases to be applied by the wrench. After releasing the wrench the wire will relax a small amount as the force is transferred to the wire lock holder. The straw bales will also relax by the small amount, but not elastically. Therefore the loss of tension due to relaxation of the straw bales may be quite significant.

A base unit that is not part of the footing can be designed to incorporate a means of anchoring the unit to the footings using masonry bolts, and means to attach skirting boards or panelling. Various refinements, including facility to incorporate service wiring behind skirting boards, have been considered but require further development.

### Use of intermediate structure units

Intermediate structure units (see page 13) may be installed between any layers of bales to enable the easy attachment of shelving, drywall cladding and other things to the wall.

There are at least three options for the basic design of intermediate structure units.

1. They may be identical to top plates if they are used only to assist in the erection of high walls (see page 40). In this case they are completely buried in the wall when the rest of the wall is built.
2. They may be designed to provide for anchorage of panelling, shelving, artwork, etc.
3. They may be designed to perform both of these functions.

## A new straw bale specification

### Why do we need a new straw bale specification? What is wrong with 'standard' straw bales?

Amateur builders and some professional builders have found straw bales to be a useful building component, but are they making do with an inferior product? What is wrong with the standard straw bales?

One could argue that the standard 350 × 450 × 900 mm bales:

- are fairly awkward to handle due to their weight and bulk, particularly when one is working at substantial height;
- vary significantly in length, which leads to an increased need for custom splitting;
- have a folded face, which usually is relatively uneven and more difficult to plaster onto, and even the cut face often has substantial ridges;
- have ends that usually have rounded and bulging features, and one end often is worse than the other; and
- have dimensions that cause the wall to be excessively thick, much thicker than necessary to achieve the usually desired insulation, and thick enough that it requires special footings, increases gross building area, and requires changes in roof overhangs and window and door depths compared with common practice.

Usually the last of these is not considered to be a fault. In fact the large wall thickness often is promoted as a benefit and considered to be aesthetically pleasing. But the option of a wall thinner than a half metre should make straw bale walls more attractive for those who do not see the large wall thickness as a benefit.

With so many faults one might ask, what do straw bale walls have going for them? So, let's check out the good points of standard straw bales as a building product.

- They are readily available in many localities throughout the world.
- The finished walls have excellent thermal insulation properties.
- They are of a size and shape that enables one to make a good stable wall, in fact a wall that has good resistance to impact and earthquake.
- Properly plastered they form very good fire barriers.
- They are environmentally friendly—a renewable, recyclable and, in some cases, a waste resource.
- They tend to be relatively cheap, because they are manufactured in-the-field using a baler, whereas nearly all other building materials undergo manufacturing in an industrial plant, at least in western countries. (It remains that a large proportion of buildings in the world are constructed of earth and straw that is dug and gathered, and not manufactured in industrial processes.)

So, although they have a lot of good points, it appears from the list of deficiencies that there remains room for substantial improvement of straw bales for use as a building material.

## The new straw bale

Let us consider potentially ideal specifications for a straw bale for building, and then consider how one might approach that ideal practically and economically. The ideal bale should overcome the deficiencies of the 'standard' straw bale. Such a product might open the use of straw bales to more applications, and probably complement, rather than replace, the standard straw bale as a building material.

The likely optimum dimensions of a new bale would be about 300 mm wide by 350 mm high by 600 mm long, with *two* cut faces. (Note that we achieved two cut faces with the standard bales by trimming the folded faces before erection—see page 23), and all dimensions should be maintained bale-to-bale within fairly tight limits, say ± 30 mm in the length and say ± 10 mm in the other dimensions. The bale would be orientated with the straws horizontal, along a 300 mm dimension.

## What would this product achieve?

It would overcome, or at least substantially reduce, all of the deficiencies identified above for the standard straw bales.

- The bales would be much lighter—about 45% of the weight—and much less bulky, and therefore easier and safer for manual handling.
- If the dimensions could be kept within the error limits there should be much less requirement for custom splitting.
- With two cut faces, and therefore no folded face, plastering would be easier.
- Squarer ends would reduce gaps between bales.
- The wall thickness would be reduced to about 130% to 140% of standard external walls, instead of 185% to 200%; a standard double-brick wall is about 270 mm and a brick veneer is about 240 mm thick. If the plaster, or other wall facing, were about 25 mm thick, the finished wall thickness using these smaller bales would be 350 mm, compared with the usual 500 mm for the usual straw bale wall. It might be feasible to reduce the bale thickness even further, but that is likely to diminish substantially the thermal insulation benefits, and to make it difficult to satisfy structural constraints.

## Can this new bale be made economically, and will anyone make it?

The cost of the new bales could be kept low only if they were produced in the field, rather than manufactured by subsequent processing, but currently the dimensions of bales are rigidly set by the baler geometry, except the bale length is adjustable from about 300 mm to more than 1 metre. However, the bale length will still tend to vary significantly due to a number of variables (straw variability across the field, varying nature of windrows, speed of tractor, etc.) and the ends will remain somewhat out of flat.

A height of 350 mm is the same as current bales. Since the length can be set to the desired 600 mm, albeit inaccurately, that leaves only the 450 mm dimension to be reduced to 300 mm, which can be done by trimming 75 mm of each face—each face rather than one only, since otherwise one of the strings would have to be removed or replaced. Trimming of the folded face is highly desirable in any case, as previously discussed, and it is also desirable to trim the cut face, because balers tend not to produce a very good trimmed surface. Seventy-five millimetres off each face is more than necessary to get an acceptably cut face, though even with 75 mm trimmed off the folded face, it still tends to retain slight folded features.

For production purposes the trimming needs to be done in the field as the bales are made, and this could be achieved with an attachment that trims the bales as they emerge from the baler. Alternatively, existing standard bales could be trimmed with a stand-alone trimming machine on-farm or even at the building site. In the latter case the cut straw could be used for garden mulch and/or composting as part of the building project. On-farm the cut straw could be either left in the field, or bagged (or otherwise collected) if there was a ready market for mulch or compost.

In the absence of trimming attachments or machines, bales can be trimmed fairly readily using a brush-cutter, but a jig is necessary to ensure the correct amount is trimmed from the bale, and the process is quite labour-intensive.

Post-processing on the building site is likely to be an attractive option for only a small number of builders; it will not gain major acceptance in the mainstream building industry. However, a trimmer attachment for a baler for on-farm use might be quite economical. One could consider a new design baler that produces a narrower bale, but that would be a high cost development option and most unlikely to be produced unless a large market could be identified. Even if a baler to produce the narrower bales was made, it would remain that the bale needs to be trimmed on both faces to produce the desirable neatly-trimmed pair of faces. On the folded face, at least 50 mm depth of trimming would be desirable to remove the worst of the folded straw.

**Fig. 21** A bale that has been partially trimmed using a jig: 75 mm is trimmed off one face using a 375 mm-high jig, then the bale is flipped and 75 mm is trimmed off the opposite face using a 300 mm-high jig.

A

B

**Fig. 22** A: Wire clips may be used to hold the strings close together to prevent them slipping off the narrower (300 mm) bale. B: The wire clip installed.

Therefore a narrower bale still would have to be at least 350 mm wide before trimming of the folded face, so the only advantage of a new baler design is likely to be less waste straw—about 22% wastage instead of 33% wastage. Of course, if the cut straw remaining in the field is advantageous agriculturally, or the cut straw is recovered and marketed, it will not be 'waste'.

When the bale is trimmed to 300 mm each string will be about 50 mm from the bale edge, which is not acceptable (usually the strings are about 125 mm from the edge). It is not feasible to change balers to bring the strings closer together, but the stability of the bale can be substantially restored by installation of clips that pull the strings close to each other at the centre of the bale. Mechanised installation of such clips would be necessary for production purposes, but manual installation is simple and quick if bales are being trimmed on-site. As well as reducing a tendency for the wires slip off the thinner bale, these clips also increase the compression of the bale. If desired two clips can be installed on each side of the bale.

### Can structurally sound walls be built with the new smaller bales?

Building a practical, structurally sound wall that is straight and without bulges or other significant defects can be achieved readily using the new method for tying down walls (see pages 59–61),but that is very difficult to achieve using the fence strainer tie-down methods.

# 9

# A wall pre-construction method

Discussion of this method of wall construction has been reserved until after introducing the smaller bale design, because, although this construction method can be used to construct walls with the standard bales, it is specially suited to the new smaller bales.

This 'pre-construction' method is effectively a method of prefabricating straw bale walls, but is designed to be done on-site. The method builds on the advantages of the new straw bale tensioning system previously described, but it is an optional extra, rather than being a required part of the system.

This pre-construction method brings the following additional benefits.

- A wall that is as flat and as straight as the horizontal surface on which it is constructed—such as the floor of the building being constructed—without any requirement for mallets or other wall straightening procedures.
- For in-fill walls, a wall that closely fits the vertical space available and avoids the necessity to insert relatively loose straw on top of the wall after the wall has been compressed.
- Elimination of the preparation of custom-length bales—splitting of bales is reduced to removal of overhang sections, and even that is eliminated by a variation on the method.
- Ability to make sections with end-assemblies that enable precision end finishing of wall sections.
- The option to 'prime' the wall surface with easy-to-apply, no waste, less mess, deep-penetrating pre-plaster treatment, thereby enabling more intimate adherence of plaster to the straw bales.
- Wall construction that is completed at floor level with no high lifting of bales—the wall may be manipulated into place by a purpose-made mechanism after it has been fully assembled except for plastering.

The wall is assembled and compressed horizontally on a flat surface in the immediate vicinity of the wall section's final location. It is then winched onto a support frame that enables it to be rotated about a horizontal axis along the centre of the wall. Using this facility the wall section can be 'prime-plastered' in an off-vertical orientation, which enables easy application of a liquid plaster mix that will penetrate the straw surface without adding greatly to the mass of the wall.

The wall section is then manipulated into its final position and locked into place for final plastering. The procedure can be used for both infill and load-bearing walls.

## Assembling the wall section

A base plate is placed on the floor with the wire lock holder loops and tensioning rods in place. Tensioning wires are attached to these and run across the floor and held temporarily in place with weights; a piece of timber can be used, but a better option for holding such springy pieces of wire temporarily in place is very small sandbags. Thereby all the compression wires are arranged in place before laying the bales. The photos (Fig. 23) show how the wire loop is attached to the base unit, and how the tensioning wire is inserted through a wire lock and through the slot in the tensioning rod, using a washer and split pin to retain the wire. At this stage the base unit can be laid flat on the floor or propped at near vertical on edge.

The bales are then simply laid on the floor edge-on, on top of the wires, ensuring that the lay of the tensioning wires is not unduly disturbed. The bales are laid in a running bond fashion, but it is not necessary to split bales at this stage. Excess bale sections can be allowed to protrude from the ends of the wall section. If intermediate structure units are to be installed, these must be placed between the desired layers as the bales are placed.

The number of layers of bales (thinking vertically, but laying horizontally) must be chosen to achieve the correct wall height after it is fully compressed. Since the amount of compression will depend on the initial bale density it is desirable to construct a test module to determine the number of layers required. A detailed procedure is provided in the box (see page 76), but the following gives the basic idea. The test module can be two bales long and need be only three bales high. Do not use a running bond structure in this test. It can be assembled in the vertical or horizontal orientation using a standard base unit and top plate. The purpose of the test is to measure the amount of compression that occurs when the design maximum tension is applied to the wires. It is then easy to calculate the free height of wall needed to produce the desired compressed height.

If the desired wall height cannot be achieved with a whole number of bales, a layer of straw biscuits can be inserted at some level. These biscuits should be removed carefully from bales and placed in a single layer. Of course, these biscuits should be laid with their straws in the same orientation as in the rest of the wall. Do not place the biscuits layer at the top or bottom of the wall; just under the top layer of bales is a good choice.

When all bales are laid—a very quick process—the top plate is placed against what will become the top of the wall and the wires are carried over the wall back to the base unit. There they are inserted through the wire locks and into the holes in the tensioning rods, and any slack in the wires is removed by pulling the wire taut with pliers only.

At this stage the positioning of the top plate and spacing of the wires should be checked. Also the bales should be pushed together reasonably firmly. Some gaps will remain at the corners of bales.

Now a light tension can be applied to the wires by turning the tensioning rods using a wrench. The wires will pull through equally on both faces of the wall and will relax only slightly before being restrained by the wire locks.

After light tension has been applied, recheck the top plate positioning, and check that the top and base plates are perpendicular to the wall surface. Then the wall can be tensioned to the full desired tension. As the tension is increased the wall height should be checked to ensure that it is compressed to fit the space available in the building structure. The tensioning rods should be left in place until the wall is in its final location, in case some re-tensioning is necessary during its final fitting.

Then the wall can be trimmed to length. The wall will have sections of bale protruding from each end. Since the wall is now in compression, these overhangs can be trimmed by cutting the strings on the end of each protruding bale, carefully removing biscuits of straw and then re-tying the strings using a butchers knot. The biscuits removed can be stored carefully for later use in adjusting wall heights. As well as removing the part of the bales that protrude by approximately half their length, if the bales generally vary significantly in length, some other bales might need to have protruding sections removed, but if the bales were properly aligned at one end, that should be necessary on only the other end of the wall section. Of course, if the desired section length is not a whole number of bales, then all bales on one end will have to be trimmed. If that is going to occur, it might be better to split some bales by the usual method before placing them in the wall, to get at least every second layer to the right wall length.

A

B

**Fig. 23** Horizontal wall construction stages. A: The base plate on the floor with tensioning wires, rods, holder loops and wire locks in place. B: The first row of bales has been placed on the tensioning wires, with the base plate propped. C and D: The rest of the bales have been placed, wires carried over to the top plate and connected with wire locks and wire loops back to the base plate, and tensioned.

C

D

**Fig. 23** continued.

## Manipulating the pre-constructed wall section for prime plastering

The compressed wall section is now of the desired height and width. Provided adequate tension has been applied to the wires the wall section can be lifted at the top of the wall using a simple lever, and a steel beam inserted under. This can then be repeated at the bottom. The rest of a clamping frame can be attached and the whole unit winched onto the main frame. The wall section can be rotated about a horizontal axis that is positioned to be along the centre of the wall.

**Fig. 24** The wall section is winched onto a special frame that enables the wall section to be rotated about a horizontal axis through the centreline of the wall. In this position it may be prime-plastered and then rolled into place after attachment to a relocation frame in the vertical orientation.

The next step, which is optional, is to 'plaster prime' both faces of the wall. The wall section should be rotated to at least 20 degrees off vertical (it may be in the horizontal position) and locked into position. In this orientation a liquid plaster can be applied to the upper face. By using a quite liquid plaster, it can be worked very easily well into the straw with almost no wastage and mess. Sufficient plaster should be applied that individual surface straws will be visible, but thinly coated in plaster, and the plaster will have penetrated well into the straw bale.

This quantity of plaster will not increase the weight of the wall by a great deal, but it will provide an excellent surface onto which the next layer of plaster can be applied after the wall has been placed into position. The wall can then be rotated and the process repeated on the other side.

The prime-plastered wall should be allowed to dry, primarily to reduce the weight by allowing the water to evaporate, but if that is of no concern, the next stage can progress immediately.

## Moving the pre-constructed wall section into position

At this stage the wall can be orientated vertically and transferred to a relocation frame that enables it to be rolled forward into its final location and locked into position by bolting it to the footings and locking the top plate to the building frame before removing the relocation frame. Final plastering of the wall section can then proceed.

At the time of writing the mechanisms for raising, rotating and manipulating the wall into position have been developed only to the stage of demonstrating the essential principles of these processes. Routine application of the horizontal wall pre-construction method therefore requires further development of some of the procedures before a fully detailed system description can be made available.

**Test module**

The following is an attempt to provide full description of a procedure that may be used to determine the number of bales, including adjusting biscuits, that will be required for the compressed wall to fit exactly the space available. Adequate results can be obtained without the following detailed analysis, but readers with a strong technical inclination might find this analysis of value.

The test module need not be a full height and full-length wall section, and the test needs to be done only once for a given batch of bales, assuming there is reasonably consistency throughout the batch. The tests can be done by assembling, in the normal vertical orientation, a three-bale high, two-bale long 'wall section' on a standard base unit. Place the bales in a square pattern, rather than in a running bond structure. Assuming one is using 600 mm long bales, with compression wires at 300 mm spacing, four compression wires can be used, the end ones being 150 mm from the section ends. If 900 mm long bales are used with compression wires at 450 mm spacing, again four compression wires may be used, the end ones being 225 mm from the section ends.

Measure the height of the unit before compression (= $H_0$)—i.e. when the wires have been sufficiently tensioned to straighten the wires, but have not begun to compress the bales. Apply the minimum tension to be used for the wall by progressively tightening each wire in small steps until the required tension is reached on all wires. As a general guide one may use a tension of 200 kgf *in each* wire if the wire spacing is 300 mm. For wider spacing of wires the tension should be increased proportionately. For example, at 450 mm spacing the tension to produce the same wall loading would be 300 kgf. These tensions will produce an ultimate wall loading of approximately 1300 kg per metre of wall length, which compares with the wall mass of about 500 kg per metre (including plaster). This is best done using a tension wrench. For tensioning rods of diameter D and wire diameter of d, the tension wrench setting, T, should be:

$$\mathbf{T = K \times (d + D)\ kgf.m}$$

where K is the tension (kgf) in each of the wires, and all linear dimensions are in metres. There is not a divisor of two, because the torque is the wire tension, K, multiplied by the radius [= (d +D)/2], *and* the torque is being applied to *two* wires, one at each end of the tensioning rod—each side of the wall. So the radius [= (d +D)/2] must be multiplied by 2K to get the torque necessary to achieve that tension, K, in *both* wires. For example, for a tension of 200 kgf in each wire, and with D = 16 mm and d = 2.5 mm, the torque, T, required would be about 3.7 kgf.m (= 27 ft.lbf).

Measure the section height (= $H_1$) after the required minimum tension ($K_1$) has been reached. Increase the tension progressively in all wires to a reasonable maximum without significantly distorting the top plate or base assembly and without breaking the wires. The breaking tension of a typical (Waratah Tyeasy®) 2.5 mm diameter (12.5 gauge) high tensile fencing wire is 550 kgf (equals 5.4 kN or 1210 lbf). The manufacturer's recommended maximum tension is 183 kgf (1.8 kN). An appropriate maximum tension ($K_2$) applied should be about 400 kgf, which is much greater than the manufacturer's recommendation, but that recommendation is for fencing. The situation here is quite different, because the actual tension will fall quite quickly from the maximum applied tension as the straw bales relax under the load. In the usual fencing situation the wire tension normally would not decrease significantly, if at all. If a wire is going to break, it will break as the tension is being applied, not at a later time, because of this relaxation of the straw bales.

The 400 kgf tension corresponds to a torque of about 7.4 kgf.m (= 53 ft.lbf) for a 16 mm diameter tensioning rod.

Re-check the height measurement ($H_2$). Calculate the fractional change in the section height from the uncompressed position to the minimum compression position and similarly from the uncompressed to the maximum compression position.

Let $C_1$ = fractional compression at the minimum compression, $K_1$
= $(H_1-H_0)/H_0$
$C_2$ = fractional compression at the maximum compression, $K_2$
= $(H_2-H_0)/H_0$

If the uncompressed bale height is $B_0$ (usually 350 mm), the desired finish wall height is W, then the number of layers of bales required to just fill that wall height at the minimum compression ($K_1$) is

$$\mathbf{N_1 = W/(B_0 \times (1-C_1))}$$

At the maximum compression value the number of bales required is

$$\mathbf{N_2 = W/(B_0 \times (1-C_2))}$$

If the whole number (integer) part of $N_1$ is not the same as the whole number part of $N_2$, then the wall may be filled with a whole number of layers of bales. For example, if $N_1 = 7.3$ and $N_2 = 6.9$, then one should choose seven layers for the wall. With that number of bales the desired wall height will be achieved at a tension between the minimum compression force, $K_1$, and the maximum, $K_2$.

If the whole number part of $N_1$ is the same as the whole number part of $N_2$, biscuits will be required, and these normally would be placed just below the top layer. The thickness of the biscuits should be chosen to make the (non-integer) number of bales between $N_1$ and $N_2$.

## A wall with bales orientated vertically and no running bond

It appears that all walls to date have been built with the bales laid flat or on their side, but never on end, and always with a running bond structure. At first sight it appears foolish to lay bales on their ends, and also foolish not to have a running bond. But the horizontal pre-construction method opens further novel possibilities for wall construction, because of the level of control and flexibility when one builds the wall as a horizontal unit on the floor. The influence of gravity only becomes a major factor after the wall section has been assembled.

How can a sound and stable wall be constructed without a running bond and with the bales on their ends, and why would anyone want to do that anyway, when there are perfectly good methods without resorting to such foolish notions? It can be done, and there are good reasons for considering these options.

First, consider a wall with the bales laid flat, but without a running bond. To compensate for the lack of integrity caused by there being no running bond, the wall can be tied and compressed horizontally as well as vertically. Such a process will not be attractive in most cases if the wall is built in the vertical orientation. However, when it is built horizontally on the floor, the process is quite straightforward.

In addition to a set of compression wires running vertically, an additional set of compression wires are installed horizontally using the same type of bearing plates on the ends of the wall. The length of the uncompressed wall is set slightly longer than the planned length of the finished wall and the uncompressed height is also set longer than the finished wall height. Both sets of wire are tensioned as previously described using the new type of tensioning system.

Figure 25 shows how this two-way tensioning system has closed nearly all of the gaps that otherwise are present between the corners of bales, without inserting any loose straw. This section of wall had about 5% variation in length of layers, and all consequent gaps were closed in the compression.

The fully tensioned wall section will have good structural integrity and can be made to fit the available space perfectly by appropriately adjusting the wire tensions; very little, if any, straw is wasted due to splitting the bales; and the ends of the wall are compressed by end plates and thereby produce an improved vertical end on the wall; the end plates can be locked to vertical building members or, in the case of abutting walls, whether at a corner or in a longer straight section, the end plates can be utilised to lock the wall sections together.

The wall section so made can be installed with the long dimension of the bales either horizontal (i.e. bales laid flat) or vertical (i.e. bales laid on end) with equal structural integrity. The bales-on-end option will produce a flatter,

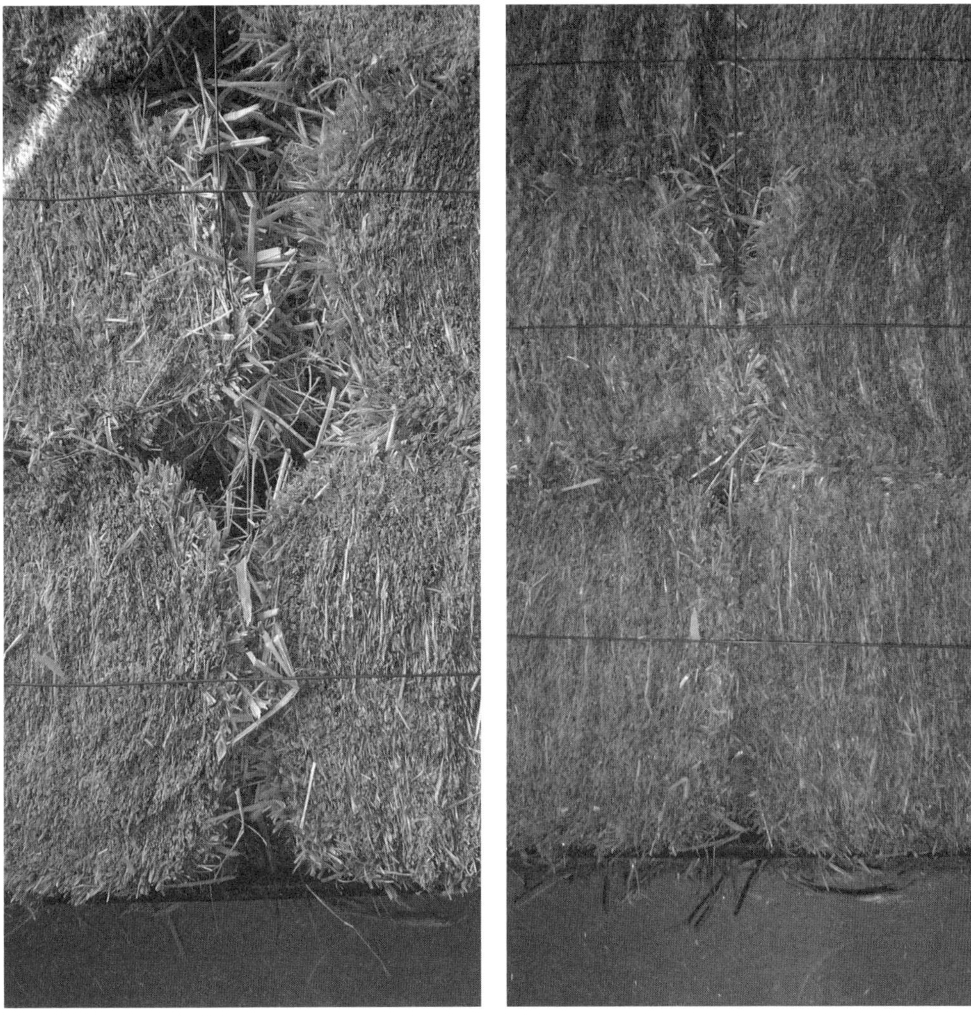

**Fig. 25** The gaps between bales (left) may be closed (right) by using tensioned transverse wires.

more even end to the wall, which makes it easy to construct a wall with a free end (i.e. no framing at the end of the wall); and the vertical compression will cause the usual holes between the ends of bales to be almost completely eliminated. If intermediate structure units are required for purposes other than aiding in building very high walls, these units may be inserted in the vertical orientation if the bales are laid on-end.

Finally, consider a wall that uses standard bales (350 × 450 × ~900 mm) with the bales orientated on-end but with the 350 mm dimension through the wall; the straw will be parallel to the face of the wall. This will produce a wall thinner than that normally produced with standard bales, but a wall thicker than one gets with the new small bales. The normal reasons why this would be dismissed are that, when the bales are laid on edge:

- vertical compression of the wall can occur essentially only by the straws bending under the compression force, (but that is somewhat a fallacy because within biscuits the straws are fairly randomly arranged); and
- it is more difficult to plaster on to straw that is lying parallel to the surface, including being more difficult to get the plaster to penetrate the bale.

However, when the horizontal pre-construction method is utilised, bales on-edge, but on-end, have the straws horizontal, so there is no impediment to strong vertical tensioning. The horizontal tensioning need only be sufficient to produce a structurally cohesive unit without making the tension high enough to significantly distort biscuits.

This procedure results in a wall that is 100 mm thinner than the common straw bale wall (350 mm instead of 450 mm before plastering); it has minimal wastage of straw; it has very good straight and flat ends; and is structurally sound. There is also no need to pre-trim the folded face before construction, because all folded faces can be hidden, provided one takes care to orientate the bales at the ends of the wall with the cut faces exposed.

However, it remains that the straws are parallel to the face of the wall, which makes plastering more difficult. The horizontal pre-construction method provides the answer yet again, because the wall can be readily prime-plastered when it is orientated off vertical, as previously described. If desired, the wall can even be set horizontally for this prime-plastering.

## The 'best' wall structure and construction method

No doubt the following is not the ultimate answer for straw bale wall construction, because improved processes, materials and methods will be devised, and more extensive testing and practical experience will be desirable before one can be too categorical, but at this stage the following appears to be a good candidate for the best straw bale wall structure and construction method.

This 'best' straw bale wall is one that: uses thin-bales-on-end, has straws through-the-wall-orientated, is horizontally-pre-constructed, is double-tensioned, and is horizontally-prime-plastered. In more detail that means:

- bales are of the new specification: 350 × 300 × (approximately) 600 mm;
- bales have two neatly cut faces;
- bales are orientated end-on with the straws horizontal through-the-wall;
- the wall is constructed horizontally using the horizontal pre-construction method;
- bales are laid in a square pattern rather than with a running bond;
- the wall unit is double-tensioned—vertically and horizontally; and
- the wall is prime-plastered in an off-vertical position.

The following are the advantages of this wall:

- bales are lightweight and easy to handle, and no high lifting of bales is required;
- the final wall thickness is not so much greater than that of other walls that it requires a major deviation from normal building practice with respect to the design of other building components;
- there is very little wastage of straw due to splitting and trimming bales, and therefore less mess of straw on the building site;
- it is very easy and quick to make the walls straight and flat without the use of large mallets and other 'persuasion' methods—the tensioning method is precise, fast, and identical on both wall faces, so repeated re-alignment of the wall is not required;
- there is almost zero tendency for walls to bend or bulge during the wall compression process;
- there is almost zero wastage of plaster during application of the first (prime-) plaster coat and there is minimal mess of spilt plaster;
- the off-vertical application of the prime-plaster enables greater plaster penetration than normal;
- the wall is thick enough to retain the high thermal resistance that is characteristic of straw bale walls;
- when used for in-fill wall construction, the wall can be made to fit exactly the space available, thereby eliminating the usual gap between the top of the wall and the building frame;
- the wall can be used as a load-bearing wall, though further research will be necessary to determine the safe limits with respect to the height and loading parameters; and
- the process is faster, because, of the above advantages, the following are time-saving: minimal, if any, bale splitting; faster bale placement; no use of mallets; wire tensioning is faster; faster prime-plaster coat; site clean-up time reduced; no filling of gaps at the top of in-fill walls or at corners of bales in the wall.

However, everything has not been perfected. It remains that the equipment for and process of manipulation of the wall section from on-the-floor to its final position requires some development, though the principle has been demonstrated.

# 10

# Waterproofing, plumbing, openings and drywall finishes

## Waterproofing wet areas

The strong weight of opinion is that very wet areas, such as shower recesses, should be kept clear of straw bale walls. Although that may be desirable, some people will not want to be so constrained, so let us consider some options.

Potentially wet areas can be plastered with a plaster made waterproof by the use of appropriate additives, of which there are many on the market. However such a plaster might still crack, and corners and areas such as wall–floor joins are particularly susceptible to leakage problems.

A waterproof membrane that can accommodate some movement is a better option. There are a number of commercial products; the more flexible are un-reinforced acrylic, polyurethane and latex formulations, which can provide a highly flexible membrane that is fully waterproof.

These products are generally designed to be used under tiles, rather than providing the wall finish. Tile adhesives, which are usually acrylic-based, will adhere well to these membranes, and the composite of a plaster on the straw, coated across the whole wet area with such a waterproof membrane, and then finished with ceramic or other appropriate tiles, should produce a fully waterproof barrier that will accommodate slight movement due to environmental change that might cause narrow cracks in the underlying plaster.

One might not want to tile all of the wet area sections of a straw bale wall. In that case the membrane may be applied over the base wall plaster and, when that has cured, it can be primed with a suitable acrylic primer that will enable a finishing plaster to adhere. There is no guarantee that this will work with all membrane/acrylic-primer/plaster combinations.

Another option is to sheet the straw bale wall with waterproof panelling prior to tiling. In this case joins in the panelling and edges/corners of the panelling should be treated with the type of flexible waterproof membrane described above. The panelling can be installed by first erecting a stud wall structure on the face of the straw bale wall, or by fixing the panelling to intermediate wall elements (see page 63).

Although these waterproofing issues are of particular importance in the case of straw bale walls, such waterproofing issues are relevant when other building materials are used. Therefore, builders/plumbers and product suppliers may be consulted for advice on standard methods.

One feature that is of particular relevance to straw bale walls is that waterproof wall panels or membranes should reach beyond the bottom level of the bales, which should be placed so their bottom is above the floor level. In such a configuration, the areas of most concern are the vertical corners and joins at the edges of panels, provided, of course, that the raised footing is itself waterproof. If that is wood rather than concrete that is integral with the rest of the footing, it should be protected with a membrane as described above.

## Plumbing

Water pipes are potential sources of dampness problems, not only due to leaks, but also due to condensation of water vapour on cold pipes. Therefore, wherever water pipes might be in contact with straw, they should be sheathed in a waterproof membrane that is also sufficiently insulating that water vapour will not condense on the outside of the membrane. If the pipe is to be straight through the wall, this can be done by inserting plastic pipes—e.g. conduit, plastic irrigation pipe, plastic drainage pipe, sewerage pipe—that will accommodate the water pipe with a reasonable fit. It is best to do this during wall construction, because it is quite difficult and time-consuming to make holes in the finished wall to enable such pipes to be fitted. Use of such plastic pipes enables the wall to be completed before this plumbing is installed.

It is highly desirable to avoid water pipes in straw bale walls as much as possible. For example, design showers so they are on internal non-straw walls. If pipes are to be incorporated into straw bale walls, at least they should be thoroughly tested for leaks before the walls are plastered.

The success of any waterproofing generally depends on the builder taking great care to follow procedures correctly, including the correct use and application of the various materials. It remains that it is better to avoid designs where very wet areas, such as shower recesses and laundry tubs, are located adjacent to straw bale walls.

## Electrical wiring

Electrical wiring and outlets must be installed in accordance with relevant building codes. Some codes will permit electrical cables to be laid on the straw face and held temporarily with wire staples until the plaster is applied over the cables, but some codes will require that the cables be run inside conduit. In the latter case the conduit can be fixed in place with wire staples. Depending on the density of the bales and whether the conduit can be run along the join between bales, it may be necessary to use a brush cutter or similar tool to cut a groove in the straw to accommodate the conduit.

At wall plug locations, metal 'plaster plates' may be held in place with wire staples until the wall is plastered. The back of power outlets should be protected from the straw with a plastic cover, so the bare ends of wires and connections cannot come into contact with the straw, in case of sparking at a faulty connection.

## Windows, doors and other openings

Straw bale wall construction around openings is least complicated when the opening extends the full height of the wall—e.g. a full-height window, or a door with window above. When an opening is inserted with bales below and/or above the opening, special measures should be taken to accommodate the vertical compression of the bales after they are placed, whilst avoiding stressing the frame around the opening.

If the opening is a whole number of bales high when compressed, the frame may be made to the height that the relevant number of bales will be after compression. Above the frame there should be a lintel that extends well beyond the opening width, and is strong enough to support the finished wall above. If the opening is wider than the tie-down wire spacing, the frame should incorporate temporary internal supports at the tie-down wire locations to prevent distortion of the frame during tie-down, and the frame should have temporary cross-braces to keep it square.

The frame will be an outer frame that is large enough to accommodate the finished frame with some packing pieces between. Thereby, small distortion of the outer frame can be accommodated during wall construction.

With the frame in place (see diagram) the wall may be tied down. The size of the frame should be quite accurately estimated from measurements of the amount of compression obtained with the straw bales being used in other full-height wall sections. If a multiple of full bale heights does not accord with the designed opening height, part-bales (biscuits) may be placed between the frame and the lintel or below the frame to position the frame as desired.

The tensioning wires can be run the full wall height (i.e. across the opening). As tension is applied the floating frame will move with the bales until the lintel above causes it to be clamped into place. After the wall section containing the opening is fully tensioned, the wires across the opening may be cut and removed, and the internal supports may be removed. There will be only slight spring-back of the bales, and the bales above the opening will be restrained between the lintel and the top plate, both of which might undergo slight distortion. The bottom of the frame may also distort slightly, but these distortions should not be significant, particularly if the lintel is designed properly to take the upper wall load.

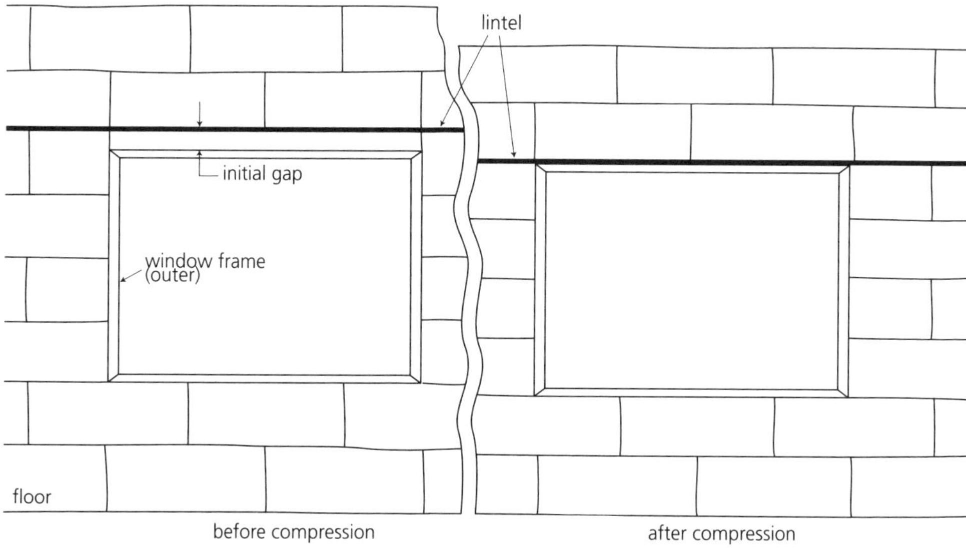


**Fig. 26** When installing windows, allowance needs to be made for the compression of the bales.

The above comments relate to wall construction in situ. Alternatively, wall sections can be constructed horizontally incorporating the opening complete with lintels and outer frame in the equivalent procedure as described above for a vertically constructed wall.

The standard use of flashing around windows to ensure that water is directed away from the interior of walls is particularly important in the case of straw bale construction. This is especially important with aluminum windows that have internal drainage paths.

## Other wall finishing methods

The vast majority of straw bale walls have been finished with a plaster—earthen, lime or cement—but plaster is not the only option. Straw bale walls can be sheeted with drywall sheet building material, internally or externally.

This is likely to suit some designers, builders and applications. Water barriers and breathable walls remain important considerations. The intermediate structure units (see page 13) provide a rigid framework onto which drywall panels may be attached with screws. These intermediate structure units may be placed at approximately 350 mm or 700 mm intervals. They must be orientated horizontally unless the bales are placed without a running bond using the double-tensioned horizontal pre-construction method.

To achieve a very flat drywall the intermediate structure units must be accurately aligned. Accurate alignment of these units is considerably easier if the new wall tensioning system is utilised, because of the precision and uniformity of tensioning that can be achieved. Of course, this presumes that the units are all accurately aligned before the tie-down wires are tensioned.

# 11

# Building codes

The Building Code of Australia (BCA) incorporates 'Performance Requirements' with which a building must comply to achieve a 'Building Solution'. A Building Solution may be achieved by satisfying 'Deemed-to-Satisfy Provisions' and/or 'Alternate Solutions'. The Deemed-to-Satisfy Provisions provide detailed descriptions of means of construction that are deemed to satisfy the Performance Requirements. Adoption of these solutions is the common procedure followed in the vast majority of building work undertaken in Australia.

The use of straw bales as building material is not specifically recognised in the BCA. Use of straw bales for building falls into the category of Alternate Solutions, which entail various options.

Some 'proven' Alternate Solutions are solutions contained in specified other guidelines, such as the *Fire Safety Guidelines* (Ed. 2001) published by the Australian Building Codes Board, but not actually forming a part of the BCA.

If one wishes to use an Alternate Solution that has not been accepted as 'proven', then it must be assessed. The methods of assessment are defined in the BCA (ref. A0.9) and include 'a Verification Method which is not listed in the BCA'. Thereby, the BCA provides a great deal more flexibility than is available in many other building codes. These other Verification Methods may include engineering calculations, technical tests and inspections, and reference to qualified experts. However, the chosen Verification Method must be acceptable to the appropriate authority, such as a local council.

Other standards contain provisions similar to the Performance Requirements provision of the BCA. For example, the Draft Australian Standard DR 02124 (a revision of AS 3740 1994), *Waterproofing of wet areas within residential buildings*, states: 'This standard shall not be interpreted as preventing the use of materials, systems or methods that meet performance criteria set out in the Standard, but are not specifically referred to herein'.

## Fire resistance

Tests undertaken by CSIRO in Sydney indicate that straw bales rendered with earthen, lime or cement render are all non-combustible in bush fire conditions using CSIRO Fire Science Laboratory's standard bush fire test conditions.* Of course, straw bales with no render will burn, but even without render fire will penetrate only slowly through a compressed straw bale wall, partly due to protection from initial charring of the surface.

However, during construction the fire hazard is very significant, because a fire can be ignited easily on the surface of a straw bale wall by nearby welding, grinding and other open flame operations. Once ignited a fire on a vertical straw bale surface will spread very rapidly up the wall. It becomes like a grass fire, but worse, because it is on a vertical face. Although it will not burn into the wall very rapidly, it will reach the ceiling in seconds and then can do a lot of damage.

It is interesting to observe that here we find yet another advantage offered by the horizontal pre-construction method. Not only does the method tend to produce less generation of waste straw on the building site, but if the wall is prime-plastered before it is put in place, although that will not fireproof the wall, it will reduce the probability of ignition if a fire source inadvertently reaches the wall surface. But this should not be employed as an excuse to be more lax with operations such as grinding in the vicinity of straw bale walls.

* Note: Australian Standard AS 3959 sets out requirements for the construction of buildings in designated bush fire prone areas.

# 12

# Garden walls and creative shapes

Garden walls can be constructed of plastered straw, but since they are not covered with a roof, particular attention should be paid to protecting the top of the wall from water entry into the straw. A simple and secure method is to cap the wall with a waterproof cover—a mini-roof—which may be metal or other roofing material, including an impervious masonry top.

A less secure alternative is to install plastic underlay sheet on the top of the wall before render is applied. The purpose of the plastic underlay is to direct away from the straw any water that enters the top of the wall through cracks that might occur over time. Although some garden walls built by this method appear to be quite sound after several years, that is a short test period. It remains desirable to use a suitable mini-roof for straw bale garden walls to ensure that they will survive for many decades.

Of course the horizontal pre-construction method can be applied readily to the construction of garden walls, though a suitable flat surface on which to construct the wall might not be so readily available.

The footings of garden walls should reach sufficiently above the surrounding ground level on both sides of the wall to prevent later problems with water. Sloping sites need to be considered carefully with respect to where water will flow and lie. Substantial weep or flow holes through the footings will be desirable in some situations. Although it is, no doubt, possible to design moisture barriers and structural features that would enable straw bale walls to be used as earth-retaining walls, the extra effort required usually would not make straw bale construction cost- and labour-effective compared with use of other materials.

The raised footings required for straw bale garden walls lend themselves to easy utilisation of the new tensioning system (see page 60), because there will be free area above ground where the tensioning tubes can be inserted in the footings at the time of laying the concrete. These tubes should extend sufficiently from the face of the footing to enable one to attach the twisted wire loop wire-lock holders.

The ends of these tubes must be kept free of concrete during pouring of the footings, and there should be means to find them easily when the footings are finished. To satisfy both of these requirements, a suitable method is to insert into each end of each pipe a plastic tube that fits the pipe fairly neatly and extends well beyond the edge of the footing. The plastic tube is withdrawn after the footings have been poured and cured, and only just before tie-down wires are installed.

## Curves, slopes and holes and other things

Straw bales lend themselves to the construction of relatively free form structures. For example the garden wall illustrated is curved and tapered on a varying slope base. The footing was laid using flexible flat galvanised iron sheet for the formwork and holes were included in the footing for tie-down wires. The first layer of bales was laid as full-thickness bales on top of the footings (with a damp course between). The other layers of bales were then laid in the fashion detailed in the diagram, so the topmost layer was composed also of full-thickness bales rather than including any biscuits of straw in that top layer. To achieve the curve of the wall, each bale was simply deformed into a curve before placing it.

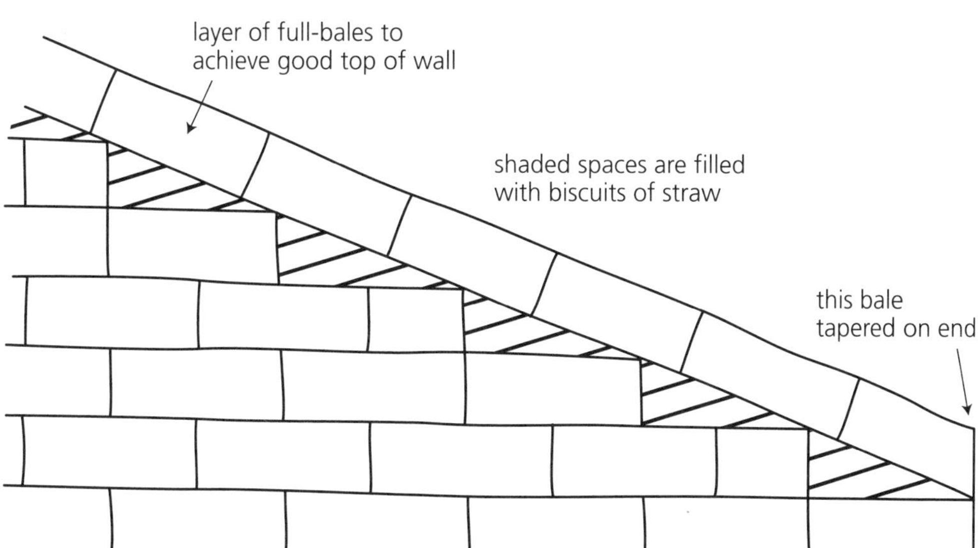


**Fig. 27** Method of laying bales in a strongly tapered wall. Note the top layer is composed of full bales laid parallel to the top of the wall. The hatched areas are filled with straw biscuits.

**Fig. 28** A curved and tapered wall.

The top plate was not pre-made, because it was not feasible to calculate the required shape of the top plate before the bales were in place. With the bales all in place two pieces of 16 mm NB steel pipe were bent to fit the curvature of the top of the wall. Then pieces of 12 mm reinforcing rod were welded between the pipes, so the curved ladder-type top plate conformed to the contour of the top of the wall. There seems to be no easy way that a suitable wooden top plate could be constructed for this application.

The wires could then be installed and the wall tied down. In this case fence strainers were used, but the new tensioning system would be a better option. There is another twist to tying down strongly tapered walls: vertically orientated wires sometimes will tend to distort the wall from the desirable shape. To overcome this, one can slope some wires or install additional wires to pull the wall into the desired shape.

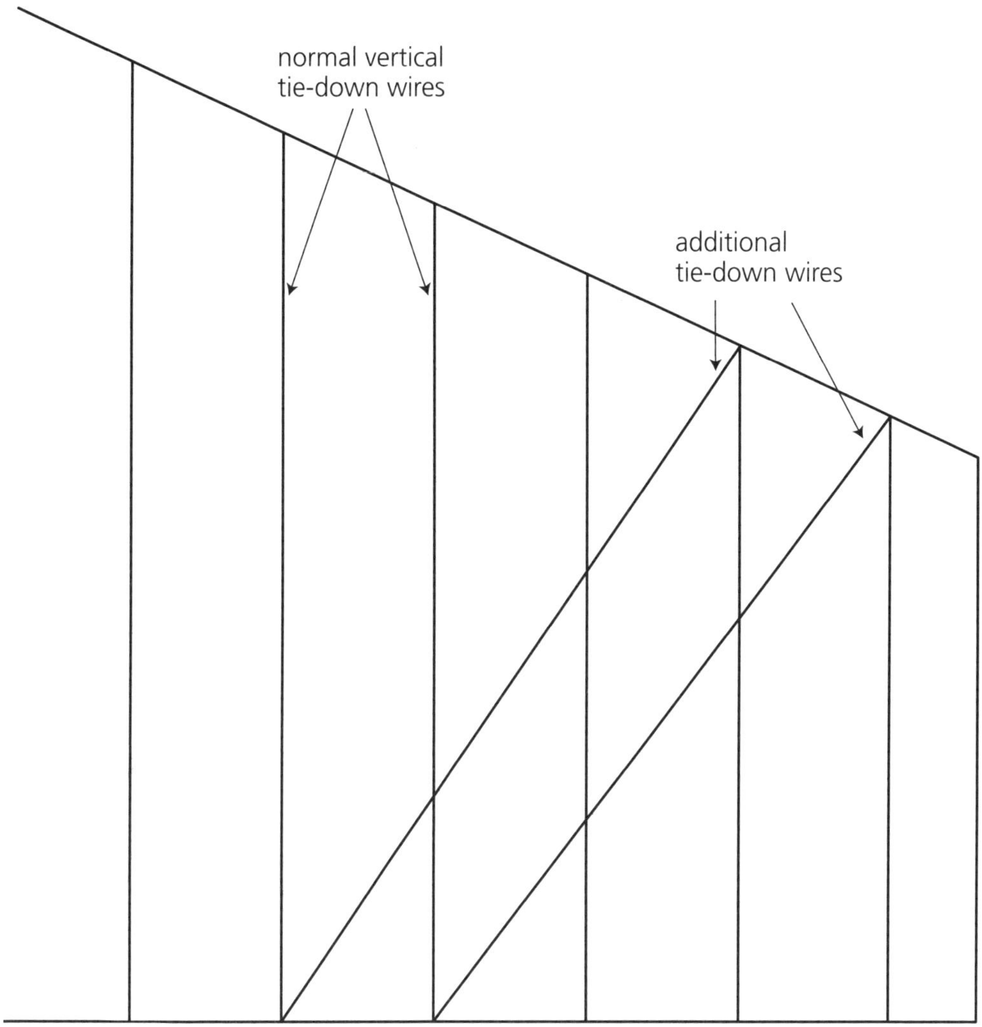


**Fig. 29** The technique of using additional slanted tie-down wires to prevent distortion of strongly tapered walls.

Bales can be shaped in any way that is consistent with structural requirements. Simple bending of bales is straightforward, and can be done to a curvature radius at least as small as 1.5 m mean radius. However one should note that the lengths of the inner and outer faces of the curved bale remain equal. Therefore when placed together in curved wall or an arch, gaps will occur and should be filled with straw. For walls with only small curvature (large radius), the gaps might be small enough that no filling is necessary.

Straw filling the gaps should be orientated in the same manner as the straw in the bales, and is best inserted as the bales are placed, rather than after all bales are placed. If excess straw is allowed to protrude from the wall, that can trimmed after all bales are placed and the wall has been tied down.

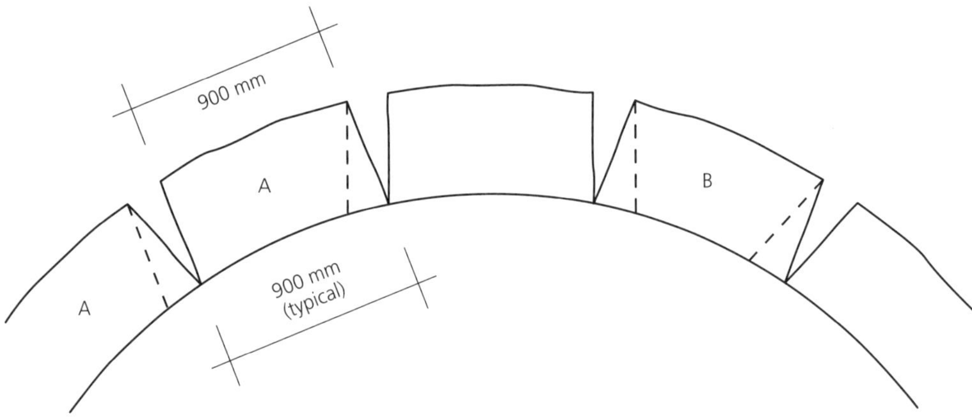


**Fig. 30** Plan of a curved wall showing how gaps will form due to the inner and outer faces of the bales being of equal length. The gaps may be eliminated by tapering the bales as indicated by the dotted lines. A and B are alternatives.

For large curvatures, such as might occur in arches, filling the gaps with straw biscuits is not very satisfactory. To eliminate the gaps the ends of the bales should be tapered before the bales are bent. Bale ends can be tapered by inserting a shortened tie and then trimming the bale end to a taper with a brush cutter. This is fairly labour intensive, and therefore should be done only if the gaps become a significant problem.

The bale can also be tapered along its length. This will require the installation of new ties if the taper is substantial. If the taper brings the face close to the tie, but does not interfere with it, it might be sufficient to pull the existing ties towards each other using the clips described for use with the new smaller straw bales (see page 64).

Even the height of a bale can be tapered, rather than using biscuits in a tapered wall, as described above. That will require the installation of new ties using a needle that is longer than the bale length. Removing the unwanted straw in this case is quite messy, because of the random straw arrangement within biscuits, so it is not very satisfactory.

Small niches in bales can be made for decorative purposes. Holes can be cut, though not very easily if it is done after the wall is built. A variety of tools may be used for shaping bales and making holes in bales. These include: brush cutters, nylon cord trimmers, hedge trimmers, simple handsaws and custom tools such as the push-pull device previously described (see pages 16–17).

Drills are not very useful, because the straw tends to wrap around the drill and can bring it to a rapid halt.

# 13

# Handling, transportation and storage

Where can you buy straw bales? A good place to start is stock and station agents. Often they will know of farmers who sell straw bales. Generally harvesting will take place in early summer, so that is the best time of year to look for straw bales. Bales can be purchased from stock feed suppliers, but normally they will be much cheaper direct from farmers, preferably direct from the field after harvesting and baling.

Before buying straw bales, check that they are reasonably tightly bound, have a generally good appearance, are not distorted, are dry and do not show signs of having previously been wet—e.g. signs of mildew or rotten straw. Check the cut faces—they should be cleanly cut without folded or loose areas.

The type of straw seems not to be very important, but straw that is composed of stiff, rather than soft, stalks and has mainly long stalks is probably better. Some suggest that the higher silica content in rice straw is desirable, though oaten straw, for example, has stiffer straws than rice straw and can produce a better face for plastering when well trimmed. One could analyse the finer points of the different straws, but there are much more significant issues to consider. Usually the straw that is most readily available is the most appropriate, because the cost, particularly the transport cost, probably will be substantially lower than for other options.

## Handling and transporting bales

Handling and transport of straw bales can be a major effort and significant expense, and it can present substantial health and safety hazards; so how and how far to transport, and who does the transporting, should be carefully considered at an early stage.

Standard straw bales are quite heavy—typically a bale weighs about 20 kg—but they can be considerably lighter or heavier, depending on how tightly they have been bound and their length. They are also an awkward size to handle. Therefore, one should be fully conversant with the theory and practice of safe manual handling before attempting to load a truck, build a wall, or do any handling of straw bales. This subject is covered exhaustively in safety publications.

Transporting straw bales is not trivial. Except perhaps for the very smallest of jobs, an experienced carrier should be engaged to collect and transport the bales, and to stack them in the storage location. An untrained person might stack bales too high on a truck, not tie them down sufficiently securely, and not drive the loaded vehicle with the care that is necessary. If a car-trailer is used to transport bales, be careful—a car trailer cannot safely carry many bales before it becomes unstable, too heavy and very difficult to properly secure the load.

## Storing straw bales

The best practice is to have the bales delivered to the site after all other preparations have been made and everything else is ready for wall construction. Thereby, most of the problems of storage can be eliminated. If the bales will be required at a time other than during the harvest season, it might be necessary to store them for some time. Check with the supplier of bales to determine their availability at different times of the year.

The bales must be kept dry at all times. Once wet through the bales will dry very slowly and are likely to rot. Do not presume that there will be time to place all of the bales in the structure before it rains—probably it will take longer than estimated, perhaps much longer! It is important to plan carefully the storage of bales before making procurement and transport arrangements, even if a very short storage period is anticipated.

If the bales can be stacked inside a building, that is ideal. If they are to remain there for any length of time (say more than a few weeks) precautions against attack by rodents should be taken. If rodents build nests in the stack of bales, they can cause substantial damage; they eat out cavities for nests and they eat the baling twine. Apart from rodent baits, an option is to lay a line of builder's lime on the floor around the edge of the stack; a line about 50 mm wide and 5 mm thick should suffice, but it should be checked from time to time. The rodents do not like to cross such a line of lime.

Bales can be stacked directly on the ground if it is acceptable to sacrifice the bottom layer of bales, because they will get wet or damp. Rather than attempt to protect that layer with plastic sheeting beneath—water will still get in—one can stack that layer edge-on (with the straws vertical). Water will not wick up the vertical straws as much as it will up a bale of horizontal straws. A layer so stacked will keep the rest of the stack dry, provided it is adequately covered.

**Fig. 31** When bales are not stored correctly they can quickly become useless for building.

After completing the stack, it must be securely covered to prevent water ingress. Assuming a place has been chosen that will not be subject to flooding, the next task is to protect the stack from rodents before it is covered. Use the line of lime method as described for a stack in a building.

The best non-building cover for the stack is a heavy-duty tarp, properly tied and loaded. However, if that is not available, plastic sheet designed for use as an underlay under concrete slabs is suitable. This should be available in 4-metre-wide rolls, 50 metres long, and is not expensive. However, if it is not properly secured, it will flap in the wind, and high winds will soon destroy the cover. The following method works.

The stack must be fully covered, right to the ground. If the plastic sheet is not large enough to cover the stack without joins, it can be joined using duct tape. Overlap the two pieces of plastic by about 200 mm and carefully seal both edges of the join along their whole length. Cover the stack so the plastic sheet extends about 500 mm beyond the bottom of the stack. Cut or fold the excess and neatly fold the main sheet as necessary to achieve a snug fit of the plastic sheet. The excess that extends onto the ground should be held down with a good layer of soil, rocks or sandbags.

The cover can still be destroyed by moderately strong winds, so the next, and final, step is to hang weights, such as old car tyres, on ropes that run over the top of the stack. These must be hung at as many places as necessary to dampen any tendency for slightly loose sections of the plastic to flap in high winds. Pay particular attention to areas where the plastic is joined with tape. Horizontal, circumferential ropes can also be used to secure the hanging weights more firmly.

Finally, the stack should be inspected regularly to ensure that the cover remains secure.

Such a stack is not very convenient for intermittent access to use some of the bales. Therefore the building program should be managed to avoid use of the bales in too many stages. The stack must be fully re-secured after bales are removed.

Clearly, if it takes this much trouble to protect the bales, they should be delivered only when everything is ready to place them in the walls. Plan ahead. The bales are just about the last thing needed.

# Index